RED GOLD

RED GOLD

THE MANAGED EXTINCTION
of the
GIANT BLUEFIN TUNA

〜〜〜

JENNIFER E. TELESCA

UNIVERSITY OF MINNESOTA PRESS
MINNEAPOLIS
LONDON

Furthermore:
a program of the J. M. Kaplan Fund

The University of Minnesota Press gratefully acknowledges support for the publication of this book from Furthermore: a program of the J. M. Kaplan Fund.

The University of Minnesota Press gratefully acknowledges the financial support provided for the publication of this book by the Offices of the Dean and Provost at the Pratt Institute.

An earlier version of chapter 5 was originally published as "Consensus for Whom? Gaming the Market for Atlantic Bluefin Tuna through the Empire of Bureaucracy," *Cambridge Journal of Anthropology* 33, no. 1 (spring 2015): 49–64.

Published by the University of Minnesota Press
111 Third Avenue South, Suite 290
Minneapolis, MN 55401-2520
http://www.upress.umn.edu

ISBN 978-1-5179-0850-8 (hc)
ISBN 978-1-5179-0851-5 (pb)

A Cataloging-in-Publication record for this book is available from the Library of Congress.

Printed in the United States of America on acid-free paper

The University of Minnesota is an equal-opportunity educator and employer.

26 25 24 23 22 21 20   10 9 8 7 6 5 4 3 2 1

To Mom and Pops,
with love from the sea

Contents

~~~

Abbreviations

~~~

AAA	American Anthropological Association
AAUW	American Association of University Women
ABC	Audit Bureau of Circulations
AES	American Ethnological Society
AESS	Association for Environmental Studies and Sciences
BBNJ	[Marine] Biological Diversity [of Areas] Beyond National Jurisdiction (UNCLOS)
BCD	bluefin catch document
BCE	before the Common Era
CCSBT	Commission for the Conservation of Southern Bluefin Tuna
CE	of the Common Era
CITES	Convention on International Trade in Endangered Species of Wild Fauna and Flora
COP	Conference of Plenipotentiaries
CPC	Contracting Party to the Convention
DPCIA	Dolphin Protection Consumer Information Act (United States)
e-BCD	electronic bluefin catch document
EC	European Community
EEZ	exclusive economic zone
EU	European Union
EUROPOL	European Union Agency for Law Enforcement Cooperation
FAO	Food and Agriculture Organization of the United Nations
GATT	General Agreement on Tariffs and Trade

GBYP	Grand Bluefin Year Programme
GDP	gross domestic product
GPS	global positioning system
IATTC	Inter-American Tropical Tuna Commission
ICCAT	International Commission for the Conservation of Atlantic Tunas
ICIJ	International Consortium of Investigative Journalists
IGFA	International Game Fish Association
IOTC	Indian Ocean Tuna Commission
IPBES	Intergovernmental Science-Policy Platform on Biodiversity and Ecosystem Services
IPCC	Intergovernmental Panel on Climate Change
IPK	Institute for Public Knowledge
IUCN	International Union for Conservation of Nature
IUU	illegal, unreported, and unregulated (fishing)
IWC	International Whaling Commission
JAL	Japan Airlines
LSA	Law and Society Association
MPA	marine protected area
MSY	maximum sustainable yield
MT	metric ton
NGO	nongovernmental organization
NIEO	New International Economic Order
NMFS	National Marine Fisheries Service, a division of NOAA (United States)
NOAA	National Oceanic and Atmospheric Administration, part of the U.S. Department of Commerce
NRDC	National Resources Defense Council
NYU	New York University
PWG	Permanent Working Group
RFB	Reconstruction Finance Bank (Japan)
RFB	regional fisheries body
RFMO	regional fisheries management organization
SCRS	Standing Committee on Research and Statistics
STACFAD	Standing Committee on Finance and Administration
TAC	total allowable catch
UK	United Kingdom

UNCLOS	United Nations Convention on the Law of the Sea
UNESCO	United Nations Educational, Scientific, and Cultural Organization
UNFCCC	United Nations Framework Convention on Climate Change
VPA	virtual population analysis (model)
WCPFC	Western and Central Pacific Fisheries Commission
WWF	World Wide Fund for Nature

Prologue

〜〜〜

The Life *and* Death *of* Bluefin Tuna

HOMAGE *to an* OCEAN GIANT

The naturalist Sir David Attenborough calls Atlantic bluefin tuna the "superfish." No wonder: she is, among other things, a fetish, a memory, an ambition, a mystery, a career, a vocation, and a rush.[1] In this book, she is also a beating muscle, a world traveler, an agenda setter, an elusive data point, a legend among anglers, a status-bearing token, and, most fatally, a piece of "red gold." The cultural biography of this boundless thing of Nature is at once a tragedy, a farce, and an awe-inspired test of the modern capacity to value life itself. The Atlantic bluefin tuna fishery is one of the first in recorded human history. Today there is anxiety about hunting the very last members of her kin.

Unease is warranted. The end of the giant bluefin tuna looms over these pages, not as a prophecy rooted in statistical urgency, so common in popular discourse, but as an invitation to relate to life anew. The fight to stem the slaughter is noble. But whatever its outcome, the dream of an ocean teaming with Atlantic bluefin tuna as ordinarily known only a few decades ago—"giants"—is over. Under the prevailing conditions of valuation, tuna will remain just another commodity for sale, unable to break free from her abstract class of being to become a singular individual in the dominant society recognized for her indispensability in webs of life.

The bluefin goes unnamed. She is not Tilikum, the deceased orca at Sea World made famous in the film *Blackfish*. Spectators paid to see this massive mammal go mad and kill his keepers as he was kept indefinitely in a chlorinated bathtub that stung and reddened his eyes just like yours and mine would have experienced.

Even so, the bluefin and the orca do share something in common. Just as Tilikum was a casualty of marine parks capitalizing on—and creating value from—the commitment to preserve a natural world facing utter destruction, the very institutions mandated to conserve the bluefin have exterminated her kin over time.[2] Regulatory regimes in the name of marine conservation have helped to realize the sixth mass extinction now under way in the new planetary epoch termed by some the Anthropocene. The institution that is the subject of this book—the International Commission for the Conservation of Atlantic Tunas (ICCAT)—formed after World War II when the slaughter of sea creatures happened on a scale and at a pace never before known on Earth. Not since the demise of the dinosaurs has the planet experienced such loss of life.

There are three kinds of bluefin tuna the world over but never do they meet as they careen through ocean currents: the Atlantic (*Thunnus thynnus*), the Pacific (*Thunnus orientalis*), and the Southern (*Thunnus maccoyii*). All three look alike. The bluefin calling the Atlantic Ocean home is the largest of all tunas in the family Scombridae, which is made up of fifteen genera and some fifty other species living across the ocean at depths where light still penetrates deep enough for photosynthesis to occur.[3] My nieces and nephews will know only by book or word of mouth that—until as recently as a few decades ago—giants twelve feet (three to four meters) in length weighing well over a ton (nine hundred kilograms) swam in schools with mates as big as they raced together on annual migrations. I tell the children to imagine a team of stampeding horses or a pack of lions on the chase below the ocean surface.

With tiny scales and eyes flush to her body—more streamlined than a torpedo—the bluefin contracts her pectoral fins into slots to generate less friction when swimming. She tears through salted water like a bullet, accelerating as if her heart were a Porsche engine. One of the fastest fish at sea, cheetah-like, exceeding most speed limits on American roadways, she clocks over fifty miles per hour (eighty kilometer per hour).[4] In fact, the bluefin can cross the entire Atlantic Ocean in forty days,[5] and somehow find the nine-mile (fourteen kilometers) stretch of the Strait of Gibraltar at its narrowest point to enter the Mediterranean Sea. She enjoys one of the longest migrations of any fish on the planet.

While the bluefin travels epic distances straight across the open ocean, she also dives to depths of three thousand feet (one thousand meters) where water is black and icy cold. Some marine scientists think she communicates with her mates through flickering light. She is endowed with a "pineal window" on top of her head, between her eyes, "as photosensitive as the retina," which, like other tunas and sharks, allows her to receive faint light as she plays, couples, and chases prey up and down the water column and in low levels of moonlight.[6]

How stunning she is. A line of small triangular finlets in electric yellow rims her upper and lower back by a tail whipping in constant motion, glinting, in contrast with the dark metallic blue on her top and the iridescent silvery white on her belly. She is camouflaged when other creatures view the depths of the sea from above, or the sun and moon from below.

She is the ocean.

But who among us knows the bluefin is warm-blooded? So entrenched are modern taxonomies that it seems normal to separate, categorize, and hierarchize members of the so-called animal kingdom in the fever to define the class of vertebrates known as "fish."[7] Defying the assumptions of the naturalists—placing her somewhere on the evolutionary scale between the typical cold-blooded fish and the seafaring mammal—the bluefin can elevate the internal temperature of her viscera, eyes, and brain up to 25 degrees higher than surrounding water. "A body temperature of 80 degrees [F or 27 degrees C]! Why they're practically mammals!" exclaims the marine scientist Barbara Block.[8] This trait explains why the muscle of the bluefin and other sushi-grade tuna such as the big eye and the yellowfin is red, not white like the canned albacore and skipjack tunas. The bluefin is prized most for the color, clarity, and fattiness of her flesh. To get a good price at market, fishers must slay her as quickly as possible so that she does not produce excessive lactic acid out of panic, and marinate herself when she dies. Connoisseurs dislike the metallic taste when she does.

Today the vast majority of bluefin tuna is destined for the fresh sushi and sashimi market. This development dates only from the 1970s with the rise of the global "sushi economy."[9] Anglers no longer discard the bluefin at docks, or sell her as cheap meat for fertilizer and pet food.[10] Gone are the days when most of her mates went to canneries. The market for raw bluefin tuna has now outpaced in profits what the canneries once made. It is not cost-effective to use filet mignon in readymade TV dinners.

The global trade in raw fish meat exacts a price all people pay, evident in international headlines. It should give us pause that a 613-pound (278-kilo) Pacific bluefin tuna with flesh the color of rubies sold at Tokyo's Tsukiji Market in January 2019 for an astounding record of U.S.$3.1 million. This tremendous sale at auction does not reflect the bluefin's everyday price. Although high prices fetched in early January are not as elevated as the rest of the trading year, extravagant bids are ways for buyers to celebrate the first day of business through great shows of wealth.[11] Theodore Bestor writes: "The New Year's holiday— the longest vacation the Tsukiji traders get—ends on January 5, the day of *hatsuni* (first freight). The arrival of the New Year's first shipments is marked by ritual, and the biggest rituals surround the auctions for the first tuna, the *hatsumaguro*."[12] Bluefin from the Pacific is always sold first on Tsukiji's opening day at the start of the New Year when fish supply is low. This practice is nationalist. The red meat of tuna when paired with white rice is a symbol of good fortune, similar to the image of the rising sun on the Japanese flag.[13]

Short supply when faced with high demand makes for a fabulous price, so the logic goes. Yet how reductive a common understanding of classical economics can be, as if it could rationalize and compress into one simple pricing calculus the extermination wrought by the inequities of capital, class, law, labor, nation, state, colonialism, and empire. It is the interplay of these forces—and the lethal absurdities of their interaction in the name of marine conservation—that this book tries to untangle.

"As with persons, the drama here lies in the uncertainties of valuation and of identity," Igor Kopytoff writes.[14] The bluefin means many things to many people, but to member states of ICCAT she signifies, above all else, the truth that economic growth through the global commodities trade is an overriding national-security interest today. This truth is inseparable from another: that acquiescence to the utilitarian logic of fisheries management under extractive capitalism depends on an alienated citizenry's implicit internalization of value commoditized in such a way that it accepts as normal the extermination of an ocean giant—rich in history, complexity, and wonder—as the cost of doing business.

A frequently cited study summarily sounds the alarm: Illegal, unreported, and unregulated (IUU) fishing has contributed to the estimate that 90 percent of large predatory fish such as the bluefin, the swordfish, and the shark have vanished from the sea, eaten or discarded as incidental catch.[15] The bluefin is not immune to the trend of collapse in planetary megafauna. So devastated is the size and number of the Atlantic bluefin tuna that, since 2011, a group of scientists has agreed to catalog her as "endangered" on the International Union for Conservation of Nature (IUCN) Red List of Threatened Species.[16]

Some fisheries scientists dispute these figures, and tinker with the estimates. By doing so, they treat the problem as if it is only a matter of exactly measuring the inventory of fish: how many are gone, at what rate, by what mode. This book moves beyond the preoccupation with actuarial accounting methods as the primary means to gauge a crisis. Instead it analyzes the rapid extermination of a former ocean giant understood not as a "tragedy of the commons."[17] I insist in chapter 2 that this tragedy finds it roots in the commodity form.[18]

Despite my reservations about cataloging an unnatural disaster by the mathematics of its rate—as if numbers alone can characterize the scope and magnitude of the loss entailed—the estimates are worth review. The "biomass" or sum of all bluefin tuna said to cluster in the eastern Atlantic peaked in 1958, and declined 74 percent in the historical period between 1957 and 2007. The bulk of this drop, 61 percent, has happened since 2000. Concomitantly, for the bluefin said to bunch in the western Atlantic, the estimate of absolute decline is calculated to be 82 percent between 1970 and 2008.[19]

None of these figures gives an account of the decline in the size of the once giant bluefin tuna. Nor do they illuminate the fate of the fishers long entwined with her capture. Nor do they shed light on the other woes impacting her lifecycle, such as warming water temperatures, ocean acidification, floating (micro)plastics, and the accumulation of toxins such as mercury from coal burning and iron mining. Nor do these estimates consider IUU fishing. Trade experts estimate that the black market in bluefin tuna was worth $4 billion in the Mediterranean alone from 1998 to 2007.[20] I heard in my travels through the ICCAT network that the dealers trading illegal drugs, arms, and migrants also trade illegal bluefin tuna. That is how profitable she has become.

Atlantic bluefin tuna has built empires through millennia. Her organized capture dates back some three thousand years when the Phoenicians, master sailor-shipbuilders that they were, first developed traps close to shore so that her kin could pass through them during annual migrations, typically from May to July. Deep in a cave on the island of Levanzo off the coast of northern Sicily, a painting of people and animals thousands of years old is anchored, at the bottom, by the distinctive teardrop shape of a giant bluefin tuna.[21] Towns throughout the Mediterranean basin formed around the traps, with towers built to spot the bluefin in clear coastal waters,[22] some protected by fortresses with artillery ready to stem pirates' plunder.[23]

The artisanal method of trapping tuna in chambers close to shore is still used today, but it is not as widespread as it once was. Where there were once hundreds of such enterprises across the Mediterranean basin, as of this writing there are only 19, operating in Italy (2), Morocco (10), Portugal (3), and Spain (4). During the 1950s, the traps represented a remarkable 46 percent of the total bluefin tuna catch and 13 percent of the total tuna catch in the Atlantic.[24] As commercial fleets adopted more "efficient" year-round fishing methods, the traps lost market favor. From 1962 to 1967, the Japanese longline fleet caught five thousand to twelve thousand tons of bluefin tuna in the southwestern Atlantic in what is now innocuously called by some fisheries scientists the "Brazilian episode." In the zone where fishers captured tropical tunas such as albacore and yellowfin (destination: canneries), the bluefin has become so "fished out" that her kin can no longer support commercial catch in the southern Atlantic Ocean today.[25]

Capture by bluefin tuna trap has changed enormously in only a few decades. One Italian ex-trap fisher remarked that rather than finance a *tonnara* and employ a workforce throughout the year, "it made sense to import [other] tuna and put it in cans. There is much money to be made. [The financiers] thought: why should I invest in the *tonnara* not knowing if I will get the returns? The ocean is not secure."[26] The speculation associated with the emerging forms of finance capital changed the fishery as the market for fresh bluefin tuna went global. Speculative finance also transformed the biography of the bluefin into one of "red gold" by entrusting the ICCAT regime with the task of securing the investment member states risked by extracting what has become in only a few decades a very volatile, high-priced fish "stock."[27]

The primary methods of bluefin tuna capture are not limited to traps and longlines, the latter of which, as its name suggests, involves baited hooks attached at intervals stretching some forty miles (sixty-four kilometers) to catch fish indiscriminately. Since the mid-twentieth century, boats called purse seiners have encircled shoals of tuna with nets by tightening a noose, bagging fish like sacks of cash on a heist. Since the mid-1990s, these vessels have towed bluefin tuna long distances at slow speed in pouches so that "tuna ranchers" may release and fatten them in cramped, underwater pens. Off the Mediterranean Coast in Spain, I observed divers, some former military men, "sacrifice" the bluefin by a bullet to the head. Gunboats stood watch over the cages twenty-four hours a day to protect the investment these creatures represented.[28] Similar to other farmed carnivorous fish, such as salmon and shrimp, aquaculture targeting bluefin tuna contributes to and exacerbates the overfishing problem. To feed the voracious appetite of the bluefin, operators must purchase sardine, mackerel, and other wild fish in bulk, thereby concentrating and extending the vicious cycle of profitable extraction to satisfy sushi lovers in rich countries without abatement.

Outside the Mediterranean basin, on the other side of the Atlantic Ocean, records are not as extensive. Some accounts suggest that Native Americans in present-day Maine killed stranded bluefin tunas in tidal pools by tomahawk before smoking them. In the nineteenth and early twentieth centuries, fishers in the north Atlantic considered her a nuisance because she shredded the nets used to target herring and other commercial fish of the day.[29] New Englanders did not eat her then. Some fishers in the western Atlantic adopted harpooning methods from the whaling industry as they hunted the giants down. In fact, records from 1910 indicate that harpooners sold bluefin tuna to oilers who boiled the fish for fuel in lamps when energy from whales—not petroleum—fired economic growth.[30] Today, harpooners electrocute the bluefin at the bow by a javelin they call a "zapper."[31]

Commercial fishers are not the only ones who wrote the cultural biography of Atlantic bluefin tuna. Recreational fishers, including "roving reporter" Ernest Hemingway, considered the bluefin a portal to a dream.[32] In February 1922, Hemingway sent a dispatch to the *Toronto Star Weekly* that reflected the growing popularity of sport fishing among the leisure class. Under the headline "At Vigo, in Spain, Is Where You Catch the Silver and Blue Tuna, the King of All Fish," Hemingway proclaimed with his distinctive masculinist bravado:

> The Spanish boatmen will take you out to fish for them for a dollar a day. There are plenty of tuna and they take the bait. It is a back-sickening, sinew-straining, man-sized job . . . But if you land a big tuna after a six-hour fight, fight him man against fish when your muscles are nauseated with the unceasing strain, and finally bring him alongside the boat, green-blue and silver in the lazy ocean, you will be purified and be able to enter unabashed into the presence of *the very elder gods* and they will make you welcome.[33]

Recreational fishers indeed "welcomed" by pole and line "the very elder gods" weighing as much as 1,500 pounds (680 kilos) at lengths of thirteen to fourteen feet (four meters).

Many other stories can be told of the once-giant Atlantic bluefin tuna. One of my own begins in the 1970s when as a child I marveled at the bluefin and the other fantastical treasures that came from the Atlantic's deep sea: the swordfish with bloated bellies, the sleek mako shark with sandpaper skin, the rainbowed mahi, and the marlin whose sail flopped down her spine like the wooden fences marking the dunes nearby. I had to tilt my head skyward to see these massive game fish through my tortoiseshell glasses in summer as they hung suspended by their tails above the docks of marinas off Montauk and Shinnecock Inlets on eastern Long Island in New York State. Dockhands would weigh them for admiring crowds, congratulating the fight of "man" the

Figure 1. Ken Fraser registers the largest, IGFA all-tackle bluefin tuna ever recorded at 1,496 pounds (679 kilos) on October 26, 1979, caught while trolling Aulds Cove, Nova Scotia, Canada. Photograph courtesy of the International Game Fish Association (IGFA).

"hunter," while the athletic build of the bluefin, now stiff and dead, would spin and sway from time to time when the wind picked up. Blood dripped down her body and splashed silently onto the wooden planks. Worth not gold but pennies per pound then, the bluefin—three or four times my size—brought trophies to winners in sport tournaments. These fish seem so distant from me now.

I have seen the marine environment degrade catastrophically before my eyes during my lifetime. This observed reality informs and motivates this book. I offer it to pay homage to a majestic sea creature I have learned to "become with."[34]

This book aims to expose the predatory—and finally exterminatory—"regime of value"[35] underpinning the practices of a regional fisheries management organization (RFMO) that, by treaty, is mandated to master and control the supply of such fish as the bluefin on the high seas in the name of marine conservation. The member states that are party to the ICCAT convention collectively insist on their right to specify, reify, quantify, measure, model, and exploit the risk associated with exporting fish treated as commodified objects in the sphere of exchange.[36] ICCAT delegates must learn to downplay, refuse, disavow, erase the splendor of the sea creatures under their remit to execute the task asked of them by international law: to fish as hard as possible under the false assumption that a country's economic growth from the sea is unlimited. Conserved here, are not Attenborough's "superfish" but the export markets of well-financed ICCAT member states and the commodity empires derived from them on the world stage.

Contrary to accounts in the news media, this book shows that ICCAT is not inept. In fact, the ICCAT apparatus has worked so well that it has organized with impressive "efficiency" the profitable extermination of a former ocean giant in just a few decades. Perfectly legal this has been. Seen in this light, ocean governance when hinged to a predatory regime of value has, in effect, helped to create the conditions for the sixth mass extinction. This geological event finds its maritime roots not only in demand from an exploding human population in need of food to eat. It is also a result of the intensity of the investment in expanding the fishing effort after World War II.[37] The slaughter in fish was just one aspect of what happened when the pace of fossil-fuel consumption and the proliferation of nuclear waste, plastic, cement, and bones from the domesticated chicken industry, among other ecological disruptions, exploded on Earth during the middle of the twentieth century. Geologists refer to this time as the Anthropocene's "Great Acceleration."

Hemingway's dispatch a century ago needs to be amended. The catch of "silver and blue tuna" no longer ushers the vast majority of people into the presence of the sublime. The news headline today is "The Very Elder Gods Are Dead."[38] To create a path that leads to transformed relations between beings—to revalue the lives that have been discounted, discarded, and destroyed—is the challenge institutions such as ICCAT present to modernizers tethered to the rewards of extractive capitalism for sea power. In the words of Rachel Carson, who saw this coming more than seventy years ago, "the black night of extinction has fallen."[39] To confront, to refute, to contest the regulatory enactments of conservation that ICCAT represents is to awaken oneself to feel, to engage, to touch, to (re)imagine—that is, to love and to respect—the sea and other nonhuman natures around us in ways that offer another way of being in this world.

Introduction

~~~

The Very Elder Gods
Become Red Gold

VALUE *on the* HIGH SEAS

Care for fish on the high seas relies on international agreements. The International Convention for the Conservation of Atlantic Tunas (ICCAT, pronounced EYE-cat) is the foremost treaty among these. Its commission is tasked to steward what has become—under its tenure—one of the planet's most prominent endangered fish: Atlantic bluefin tuna. The ruby red meat of the bluefin is the most coveted of all fish in today's "sushi economy."[1] Industry insiders dub her "red gold" for the exorbitant price her flesh commands on the global market. Yet when the ICCAT convention entered into force in 1969, the bluefin met her fate at the cannery, if not the taxidermist after a sport tournament. Then fishers called her "giant" as she raced in packs like wild horses across the open ocean. Similar to the swordfish and the shark, likewise depleted, the bluefin has crashed in size and number while under ICCAT's custodianship for a half century. What is IC-CAT achieving, if not its advertised purpose to conserve sea creatures under threat?

For millennia the bluefin of the Atlantic and her cousins of the Pacific have been entangled in the lives of many peoples. This past century the bluefin has helped to transform ICCAT into a powerful international organization regulating the catch of creatures that traverse the waters of the Atlantic basin beyond the jurisdiction of any one nation-state. At the same time, the authority ICCAT has accrued in ocean governance has helped to transform the bluefin into "red gold." Their coproduction is anchored in the aspiration for what this book calls "commodity empires," detailed in chapter 1. A commodity empire is not a territory but a practice of the global historical imagination through which claims to control the ocean materialize in such activities as the extractive monetization of commercial fish. Sir Walter Raleigh's observation some four hundred years ago still holds. Whoever controls the ocean—which today includes its fish, its oil, its gas, its wind, its sea lanes, its undersea cables, even its genetic sequences—becomes the empire of the day. Had Raleigh lived in this age he might very well have found his "city of gold" inside the labyrinthine negotiations, trade-offs, and (geo)politics coursing through the ICCAT apparatus.

The general coherence in ICCAT's handling of the Atlantic bluefin tuna as "red gold" indicates that this particular life-form is more desirable than other traded fish on the global market. The bluefin is now marked sociologically, economically, linguistically[2] as "glamour fish," as "resource," as—in fisheries parlance—quantifiable "stock."[3] These terms expose the dominant "regime of value"[4] omnipresent and normalized in institutional practices alienated from the lifeworld of sea creatures. In ICCAT's regulatory zone, nonhuman animals are above all else commodified objects inventoried by their "population."[5] They are not individual subjects entitled to partake in the survival of their kin. The bluefin more than any other creature in ICCAT's convention area provides this model for fisheries management. That model is now commonplace and routinized, legal and ordinary to the conduct of ocean governance across the globe today.

It has not always been this way.

The pattern of treating fish as "biological assets"[6] has as its consequence a time now known as the sixth mass extinction.[7] Not since dinosaurs roamed this Earth has the planet experienced this much loss of life.[8] Troubling is not only the loss of this creature or that one. It is the cascading effects and ruptures to the interdependencies of life important to all beings. "Extinction," in short, "is a multispecies event."[9] The planet's sixth mass extinction is characteristic of a proposed new epoch

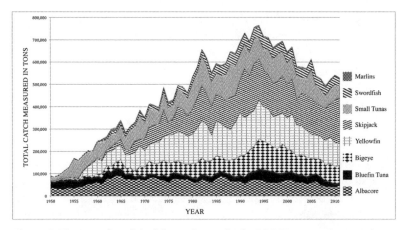

Figure 2. The rate of catch in fish accelerates in the ICCAT convention area in the period 1950 to 2012.

of geological time: the Anthropocene. The earth-science textbooks I knew in elementary school will need to be amended. The last twelve thousand years of the Holocene have come undone. Some writers contest the Anthropocene's name, and emphasize the enormous political implications of designating when this time began.[10] This book cannot possibly do justice to these important debates. Instead, it draws attention to another moment—the middle of the twentieth century—when scholars say the Anthropocene swung into the stage they call the "Great Acceleration."[11] In the immediate aftermath of World War II, a host of tremendous "socioeconomic" and "earth-system" "trends"—nuclear waste, fertilizer consumption, human population, large dams, ocean acidification—spiked across the globe with such vigor from industrialization that the course of planetary life profoundly changed. Marine fish catches is yet another of these trends.[12] Figure 2 shows the surge in the commercial catch of high-seas fish in the Atlantic Ocean after 1950. These fish are the ones that have been under ICCAT's remit since the institution formed in the late 1960s.[13]

I have written this book to expose—as exterminatory—the limits of a predatory regime of value that by treaty continues to regulate the capture of living beings as biological assets. This book makes three interrelated points. First, contrary to portrayals in the news media, explored in chapter 3, ICCAT is not inept. This book emphasizes the demonstrable fact that ICCAT member states have faithfully executed the task assigned them by international law: to fish as hard as possible so that national economies could grow. Conserved here have been not the bluefin or the ecosystem but the future end commodities supplying the export markets of ICCAT member states. The bluefin as biological asset has conferred status upon constituents best positioned to exploit "red gold" on the world stage.

This point presupposes that policy makers in ocean governance have defined "sustainability" conceptually within, and tailored it to, the national project of economic growth derived from extracting what seem like unlimited "natural resources" from the sea.[14] In the regulatory zone of fisheries management, marine conservation does not mean ensuring an ocean full of fish in perpetuity for the benefit of all beings on this planet. The ICCAT regime has not "saved" the bluefin to preserve an ocean titan for her place in the ecosystem to be revered by future generations. The ICCAT regime has instead preserved the power, the prestige, the profit the bluefin bestowed upon the delegates invested in exploiting risk when "harvesting" fish in a now global sushi economy. ICCAT member states were responsible for regulating the fishing industry. Their remit was socioeconomic, not environmental.[15]

Second, for the ICCAT regime to referee economic growth for its member states, delegates must learn to disavow the sacredness of the bluefin as a "very elder god," discussed in the Prologue of this book.[16] In the struggle to bound and impose fixity upon inherently uncontrollable sea creatures, the institution entrusted to steward the bluefin has systematically suppressed and ignored her awesome qualities. Bureaucratic protocols reduced nonhuman animals to a "class of being,"[17] represented in language, models, and metrics that together rationalized the distance—and extended the alienation—between modernizing "man" and (over)exploited fish.[18] Far from the expansive blue horizon of the high seas, ICCAT delegates did not call the bluefin a former ocean giant, let alone a tiger of the ocean, but named her in commodified form: "products" "divisible by units" to be "processed" and her kin "transshipped." Taken for granted and pervasive, this language resounded on the floor of the ICCAT meetings I attended and appeared in the mass of official ICCAT documents swollen with the techno-speak I reviewed.

Third, under the prevailing conditions of valuation, the techno-scientific institutions entrusted to conserve the bluefin have become central to her extermination. Regimes such as ICCAT have lured and abetted, conditioned and accelerated the extermination of sea creatures by exerting power over life through the banal administration of commodity empires. The ICCAT member states mandated to conserve the bluefin have not only overseen her slaughter. They have provoked it. Given the precipitous decline of big fish worldwide since World War II, regulatory regimes under the thumb of finance capital must be seen as collaborative agents in accelerating the planetary development of the sixth mass extinction.[19]

The regulated demise of the once giant Atlantic bluefin tuna lays bare the deep contradiction of extractive capitalism for the amplification of empire on the high seas. The paradoxical mandate to hunt and protect, to exploit and conserve nonhuman animals on a global scale crumbles under the weight of its own inconsistencies. Beneficiaries congratulated themselves for exercising just enough regulatory restraint to yield profit from the "efficient" extraction of commercial fish. Like the alcoholic addicted to her drink, the technocrats husbanding the bluefin returned again and again to a cocktail of law, economy, state, and fisheries science. The promise, the conceit, the self-certainty of "progress" in the halls of ocean governance intoxicated them when drunk on seemingly neutral statistics and transformative technological fixes. Sometimes they allowed themselves to be moved by the temperance movements of environmentalists in civil society. Yet a hangover is long overdue. It must force a long hard stare at the institutions high modernizers have built and citizens enabled for the accumulation of wealth organized around the social life of capital.

Taken together these points support the book's main argument: care for sea creatures must entail the rejection of a hegemonic regime of value that eradicates entire life-forms for the sake of commodity empires. That regime must be dislodged and replaced by something else. This will not happen, this book contends, until people recognize that the goal of supranational regulation cannot now be to "save" such fish as the bluefin from a creeping, labored, ominous extermination. It is already too late for that. Global elites can no longer mourn the loss of what they have systematically destroyed through the instrumentalization of law and fisheries science.[20] While the ICCAT "game" may be over for giant tuna, responses to the Anthropocene no matter its name or date are not. A damaged planet signals damaged relationships. The goal must be to cultivate opportunities for creatures to meet, in Donna Haraway's famous formulation, where beings encounter one another and engage—with curiosity, in mutuality, without aspiring for control, free from a return to a fictive Nature—so that kin such as the bluefin are not only "saved" but respected, her mysteries embraced and widely shared. The stakes could not be greater in a time of planetary extinction.

Value Matters

Rather than focus on how deteriorated the ocean has become and how threatened fish "stocks" are—these topics have been handled very well by other authors[21]—this book takes a critical approach to the practices of an intergovernmental organization instrumental in producing these deleterious effects.[22] It attempts to dismantle rather than demystify the regime of value that undergirds institutions capable of shaping in exterminatory ways the trajectory of life on this Earth.[23] I do so with the hope that more pathological episodes could be averted, not through more of the same salvage conservation but through a basic mode of engagement respectful of all beings, both human and nonhuman. This should be the floor, not the ceiling.

The crisis represented by an ocean that is said to have rapidly lost 90 percent of big fish[24] finds its roots in exterminatory regimes of value. Value cannot be reduced to an economistic paradigm consumed with price, scarcity, supply, demand, rent, fixed income, surplus capital, and the like.[25] Value in an anthropological sense is concerned with meaning making, with modes of relating to other living beings based on a shared set of standards ordering "the world of things on the pattern of the structure that prevails in the social world of its people." Made in this—a capitalist—world are not just objects. People are made in this capitalist world too.[26] The social acceptance of greed and the normalization of competition, for example, are not timeless, universal traits hardwired into what it means to be "human," let alone "animal."[27]

The bluefin's commodity state is not pre-given. It is only one of many possible modalities of relating to fish. "Other kinds of things—

like heirlooms, or children, or sacred places, for example—are typically not classified in this way," Lindsay French writes. "They have a different kind of value and, under ordinary circumstances, are not for sale."[28] A theory of value that underlies any system of classification is a struggle over which system will carry the day. How to impose this system—and not that one—is always a question of "politico-cultural categorization."[29] The ICCAT regime asserts its authority in the service of commodity empires when the bluefin becomes "red gold," "precisely by insisting on its right to singularize" and reduce individual fish to a class of precious objects for elite consumption.[30] At present the bluefin is only available for purchase in overdeveloped capitalist economies with infrastructures robust enough to fast deliver across the globe a delicacy chilled to cater to bourgeois taste.

This book understands a commodity not so much as an "object," "thing," or "delicacy" but as a "situation," which opens up space to consider exchangeability in the realm of the social.[31] The creation and maintenance of value is a political, social, and cultural question developed in complex and contingent histories. It is not simply an economic one. Value is informed by various standards, categories of understanding, and social imaginaries worked out practically by the various parties participating in or contesting the exchange of Atlantic bluefin tuna in its commodity form.[32] ICCAT is a zone, a context, a condition representative of the ways in which high modernizers invested in the ideas of "progress" commit to and validate competing values in society.[33] It would be missing the point entirely to conclude that ICCAT is a camouflage for brute economic (self-)interest or a conspiracy hatched by sinister elites, because what is culturally significant here is the fact that there is a compulsion for ICCAT and its affiliates to say otherwise, to defend themselves, against this charge.[34]

Of course, the bluefin is much more than a "thing." She is a living being. Yet she is above all else "animal capital" in ICCAT's regulatory zone. This situation reflects the ascendant "anthropocentric order of capital" in the "pursuit of new markets."[35] Nicole Shukin expertly describes what she calls "animal capital" in ways that correspond with the treatment of bluefin as "red gold": "In the ring of *animal capital* can be heard a real threat of totality posed by the global hegemony of capital . . . [T]he two words are supposed to sound almost, but not quite, the same." The dissonance this phrase evokes in fact stems from the very paradox implied by their pairing. The symmetry or "perfect tautology of nature and capital" can never be fully realized and secured in an institution such as ICCAT, try as it may, when faced with the variability in and the movement of nonhuman animal remains for exchange on the global market.[36] The bluefin responds.[37] She does not reproduce ad infinitum. She is not a widget, nor is she quicksilver capital or a cog in a well-oiled machine, even though she may be represented in these ways. "Natural capitalism" is a "social fantasy" dodging the practices of global empire and the politics of wealth accumulation through nonhuman animal beings.[38] By celebrating the entrepreneurial spirit for the making of ever more assets, proponents of this dogma think profit rationales do not need to be sacrificed and systemic failures addressed in the pursuit of finite, commodified natures.

A predatory regime of value institutionalized through the state's scientific management of fish "populations" corresponds with what Michel Foucault calls biopower.[39] Over time, the first two chapters of this book show, "the biopolitical context was progressively put at the service of capitalist accumulation."[40] That the bluefin now confronts the reader more often than not as "red gold," as export for elite consumption, already as meat and not as mystery, a fait accompli, is the trouble here.[41] Regulatory zones such as ICCAT are sites where a speciesist worldview is ordered, rationalized, normalized, institutionalized. Many of the ICCAT delegates I met assumed that marine conservation was purposed only for people, full stop, as if freedom could be finally found once unleashed from the ties that bind people to this Earth. But global elites cannot go it alone and maintain the planetary system for themselves without acknowledging the interdependency of life. How this mode of relating to living beings found one of its expressions in ICCAT is the subject to which I now turn.

The Emergence of ICCAT's Regime of Value

Although pilfering the seas is not new to the twentieth century,[42] the decline in the size and number of bluefin tuna—and of marine life more generally—quickened pace and intensified during the Great Acceleration.[43] The 1950s saw petroleum-powered fleets from a handful of industrialized countries span the globe to extract fish in new and distant waters. Before then, the high seas had been largely inaccessible to fishers. "Vast, rough, and far from coasts, fishers did not have the vessels or the gear to exploit these areas" systematically.[44] By the mid-twentieth century, times had changed. Rich countries subsidized their fleets with cheap fossil fuel and adopted technologies so "efficient"—radar, sonar, global positioning systems (GPS), onboard freezers, steel hooks, polymer line disguising baited lures—they gutted the sea of big fish.[45]

To coordinate and rationalize the extraction of fish traveling far and wide across national jurisdictions on the high seas—which today cover about two-thirds of the ocean or half the surface of the planet—a supranational regulatory body emerged in the colonialists' old stomping grounds of the Atlantic basin. A year after the photograph *Earthrise*—which showed the blue planet in color from space for the very first time in history—appeared in 1968, a small group of nation-states brought into force a treaty authorized to conserve, in its words, "tuna and tuna-like fishes." Under the auspices of the Food and Agriculture Organization (FAO) of the United Nations, the International Convention for the Conservation of Atlantic Tunas established a commission to carry out its mandate in 1969. For a half century, ICCAT has assumed the mantle of conservation by serving as the custodian of marine life on the high seas of the Atlantic Ocean and what are called by treaty its "adjacent seas," including the Mediterranean. The ICCAT secretariat is based in the landlocked capital of Madrid, from where the monarchy of Spain once directed its Armada.

ICCAT's predatory strategies of extraction rely on norms adopted after World War II when the United Nations Convention on the Law of the Sea (UNCLOS I) codified the juridical formula of "maximum sustainable yield" (MSY) in 1958, discussed in chapter 1. In accordance with international law, the text of the ICCAT convention stipulates that member states "permit the maximum sustainable catch" to "manage" what experts call "abundance" (biomass).[46]

But the boundaries of ICCAT's official remit to "manage" the "abundance" of fish on the high seas leaked from time to time. As one high-ranking ICCAT delegate expressed to me, fish were not the only "stocks" negotiated at annual commission meetings when delegates assembled to decide the rules of global trade. Also on the table—although

Figure 3. The stage is set for the banal administration of commodity empires at the annual meeting of the ICCAT commission, pictured here in Istanbul, Turkey, in November 2011.

never discussed on the floor while meetings were in session—were other goods requiring coordination for their smooth (or stymied) flow across national borders. I quote him: "It's not just a 'game' you play here . . . You also play the 'game' in other fora [like the World Trade Organization (WTO)] . . . You take our fish, we don't give you bananas or coffee."[47] Another high-ranking ICCAT delegate told me that on the negotiating table were not only fish and bananas and coffee but also drugs and people. At the time of my fieldwork, the United States worried about narcotics and European Union migrants, especially from North Africa and Asia's western edge. "Almost all those arriving [to Europe] came across the Mediterranean or the Aegean Sea," Reuters reports, or nearly one million people in 2015 alone.[48]

Rights to fish were thus "chips" that delegations played in the broader political "game" of wrestling for control over borders and "natural resource" markets.[49] Oral pressure offstage at ICCAT had the potential to impact policies and revenues from other sectors of the economy "as well as requests for foreign aid, especially from Japan."[50] Money, aid, and access to commodity markets were weapons of cooperation just as they were of coercion, depending on the negotiating context. Despite its high-modernist trappings, ICCAT was not far from the networks of patronage long documented by anthropologists in colonial settings.

A mafia, a "cartel,"[51] a "global Ponzi scheme"[52]—these are among the descriptions of a global fisheries trade considered out of administrative control by some writers. The ICCAT regime was not untouched by such dealings as it sought to professionalize the commercial fish trade through its association. "A fisherman is a thief," an industry representative bluntly expressed to me at an ICCAT meeting in 2012. "He robs from the oceans. And what a thief hates more than anything else is to have someone steal from him." It was in this spirit that well-financed ICCAT member states labored hard to ensure honor among thieves, at least for the delegations profiting from the traffic in commercial fish.

To predict, regularize, and legally constitute export markets in high-seas fish, delegates participating in the ICCAT regime think as modernizers do and perform their mandate: harness commercially important, inherently mobile things of Nature—even domesticate them—by eliminating the nomadic and the untamed to enable capital to accumulate systematically. In this complex zone of high modernity—*where to be fully "civilized" is to be removed from Nature*—the ICCAT regime attempts to contain and impose order over wild creatures by bundling public goods in the Enlightenment tradition. These goods are admirable: the advancement of knowledge through fisheries science, the use of reasoned debate as a check on power, and the protection of the rights of individual nation-states, to name a few. ICCAT ties these principles to its institutional framework and guarantees their survival by legal contract. But these principles, as sensible as they are, have become sutured to a commodified regime of value. In the ICCAT zone, delegates are obliged by treaty to predict the movement of ambulatory sea creatures to secure "biowealth," or the national property of member states participating in the ICCAT regime via the catch quota system.[53] ICCAT member states multiplied, concentrated, and distributed property rights to fish based on the hubris that experts can master and control the ocean.

In the field while conducting research for this book, I felt that

ICCAT was on occasion like an inescapable vortex that, over time, choreographed a slow-motion catastrophe; or like a corporate boardroom where executives pursued strategies for exploitation, their decision makers few; or like a complicated board game of Risk where "information [was] poor, scarce, maldistributed, inefficiently communicated and," not least, "intensely valued."[54] ICCAT delegates must learn the norms and values of fisheries management to participate in regulatory action. Many of them were convinced of the necessity of their secular models of "progress." For some of them, ICCAT served its purpose by creating an environment in which industry could compete, work, and function through the rules states adopted to govern commercial activity. Others papered over the cracks and carried on machinelike to perform their duties. Others expressed to me their frustrations about a technoscientific, legal infrastructure changing at a glacial pace in the face of rapid biodiversity loss.

This book offers an ethnographically grounded, historically informed account attempting to show how ICCAT made sense to itself in light of mass extinction. This powerful supranational organization was invested in, set in motion, and tried to effect continually a universalizing project of "one-worlding."[55] I try to trouble the narratives of "progress," not only to illuminate the dynamics of power and domination in a regulatory zone radar-locked on profiteering from the trade in commercial fish. I want to disturb the central categories of the "civilizing" project itself: "the idea of universality, the fixity of meaning, the coherence of the subject, and, of course," as Paul Gilroy writes, "the foundational ethnocentrism in which these all tend to be anchored."[56] Kris Olds and Nigel Thrift put it this way: "It is a constant in the history of capitalism that what there is to know about the conduct of business is surrounded by a garland of institutions that not only impart that knowledge but attempt to codify and improve upon it, so producing new forms of conduct."[57] To expose that conduct and its corresponding regime of value is to begin to find a place outside it from which to know this regulatory world differently or to resist directly the brutal effects of its own engineering.

Who Guards the Guardians on the High Seas?

I now turn to a brief discussion of global governance to better situate ICCAT in a world much broader than it.[58] The control and coordination of the global commodities trade through formal, legal associations is not particular to ICCAT. Emerging in earnest in the 1970s, the appearance of mechanisms to channel the flow of bauxite, iron ore, and mercury dovetailed with the creation of "informal cartels for lead, phosphate, uranium and other metals," in part driven by the uncertain economic context of the 1974–75 recession in which the "sudden, sharp, and spectacular" fall of commodity prices contrasted with their rise a few years earlier.[59] Supranational regulatory agreements, it seemed to their makers, stabilized markets.

Similar to hundreds, if not thousands, of international agreements exploding in number over the past few decades, ICCAT is a local manifestation of the intensifying reach in the forms of transgovernmental rule and the bureaucracy devised to address interdependence in an age of globalization.[60] Globally governing institutions extend the architecture of the Westphalian state system adopted in the mid-seventeenth century, which, in principle, upholds each nation-state as the sovereign of its territory and an equal among fellows in international law, no matter its size or status. Global governance in the twenty-first century is entrusted to regulate a wide range of goods and functions requiring diplomatic cooperation between states. It signifies the need for global elites to find a space to do the regulatory work of cross-border security, trade, banking and finance, telecommunications, labor standards, environmental protection, and the like at a time when the imperial and the fully postimperial are inadequate descriptors of the current world. In the establishment of supranational systems of regulation, some like ICCAT have been formed by treaty while others by informal intergovernmental networks of cooperation.[61] Commodities such as sugar, pepper, tea, coffee, and olive oil are not untouched by ultrastate or global rules of trade.[62]

Although international regulatory agreements are not new, the scale, speed, pace, and density of their establishment over the last half century—and the scope and substance of their mandates—are.[63] As of 2005, the number of formal agreements associated with global governance totaled 245, although other counts with more generous criteria suggest that their number has reached an astounding 7,306. This count is some fifteen years old.[64] Within the narrower field of ocean governance, as of this writing, there are fifty-three regional fisheries bodies (RFBs) alone.[65] Five are dedicated to regulating the supply of tunas and like fish on the high seas.[66] ICCAT is the most important of these.

Consider, as the legal scholar Sabino Cassese does, that "there are more of such organizations than there are states" belonging to the United Nations, currently at 193 members. Also important is the fact that the participation of public authorities representing home states in these institutions has tripled.[67] Increased workloads create bureaucratic, personnel, financial, and linguistic demands on all countries as they send delegates to talk, negotiate, translate, and decide in meetings the legal parameters of globalized trade and functions under conditions of tightening economic concentration. The extent to which global governance tends toward authoritarianism, fragmentation, and the erosion of deliberative rule is a concern.[68]

Unlike domestic government, global governance does not rely on police powers to achieve regulatory compliance. Without the stick of a navy, the ICCAT regime governs in a world of governments.[69] The ICCAT is not purposed to enforce rules with teeth through top-down, command-and-control measures, as domestic law would have it. The state under international law becomes not irrelevant but disaggregated,[70] its sovereignty derived not from its independence but from its interdependence.[71] Good standing in regulatory agreements is an expression of what Antonia and Abram Chayes call the "new sovereignty," a theme discussed in chapter 2.[72] Cracks in this managerial model of international lawmaking nonetheless peal like thunder throughout the book.

How to govern goods in a world of governments speaks to what Karl Polanyi called decades ago the "great transformation." As trade became linked with peace to mitigate the egregious plunder of the pirate, it grew to depend on an international monetary system that "could not function in a general war." Polanyi writes: "The balance-of-power system . . . could not by itself ensure peace. This was done by international finance, the very existence of which embodied the principle of the new dependence of trade upon peace."[73] To extrapolate from Luc Boltanski and Eve Chiapello, ICCAT is one circuit in a larger constellation of many, which, at a rudimentary level, facilitates the *"imperative to unlimited accumulation of capital by formally peaceful means."*[74] To minimize the threat of war through market economies now globally connected, ICCAT has become a space for member states to handle practically and decide jointly how to share what is left of the spoils on the high seas.

That the state and the private sector came together at all through ICCAT was by design. Their "interpenetration and contact . . . is an essential art of the structure."[75] ICCAT did not comport to some abstract fantasy of the "free market." As agents of finance capital, ICCAT member states worked out their business entanglements in face-to-face, social interactions. To put it simply, the capitalists need the state to play an active role in regulating markets, just as the state needs the capitalists to fund militaries, social programs, and the like. Their commingling is inherently (geo)political, and, in fact, "cannot be reduced to some kind of technical or administrative function."[76]

"In the end, ICCAT is a better system than a military option. By gunboats is how it used to be decided," said a fisher to me after years in the tuna business in November 2010. While armed aggression as the primary means to settle scores through fish may have dissipated, violence still haunts ICCAT's decision-making process. Over the past half century, ICCAT member states have administered the no less determined extermination of sea creatures by extending the administrative state's "project of simplification."[77] This project reduces the complexity of the bluefin to an abstract "stock" in "red gold," and favors "efficiency" and "returns on investment" at the expense of natures treated as separate from and controllable by "man."[78] Although this artifice is fragile, vulnerable, and jerry-built, its material impact on oceanic life-forms will continue to be potent, unrelenting, and decisive in the absence of another regime of value.

"ICCAT Rules the Atlantic"

"I have heard it said that ICCAT rules the Atlantic," declared a representative from the Food and Agriculture Organization (FAO) of the United Nations on the floor of a commission meeting in 2012. What, how, and for whom, exactly, does ICCAT "rule"?

Between the years 2006 and 2010, bluefin tuna "harvested" in IC-CAT's convention area accounted for 47 percent of all bluefin tuna exports the world over. ICCAT member states control nearly half of the supply in the world's most expensive tuna fish. Yet "ICCAT rules the Atlantic" not simply because the financial stakes in the bluefin are greatest there. ICCAT derives its authority, in part, from the broad representation of participating countries in regions well beyond the Atlantic basin.[79] The more countries that signed the ICCAT convention, the more influence the institution accrued; the more member states that participated, the more likely their stake in the institution's survival.[80] The ICCAT regime encourages countries to sign on. It has been successful in this regard.[81] The number of countries that have signed the ICCAT convention has more than doubled since the mid-1990s (Appendix A). Intergovernmental organizations such as ICCAT "enlarge the realm of consensuses" supporting their own power.[82]

Nearly half of the world's countries now participate in ICCAT as member states. ICCAT is the largest and most representative of all tuna regional fisheries management organizations (RFMOs). As of this writing, there were fifty-three "Contracting Parties" to the ICCAT convention (CPCs), although the European Union with its twenty-eight countries is one.[83] Not all ICCAT delegations of state participated in equal measure. Contracting Parties sending the most representatives during my fieldwork included the European Union, Japan, Morocco, and the United States. Not coincidentally, these were the same Contracting Parties claiming the lion's share of the Atlantic bluefin tuna quota (Appendix B). By contrast, poor coastal states sent just one delegate, if any at all. Whether "capacity-building measures" level the playing field or (re)entrench old dependencies between the colonizer and the colonized is a debatable question.[84]

According to a basic premise of international law, nation-states volunteer to enter the ICCAT treaty for the provision of goods, as if by private contract. The legalese designating member states as Contracting Parties suggests that they "spend their time renegotiating their contracts, that is to say, interacting and exchanging information as it is produced."[85] As ICCAT member states (re)wrote regulations about how to trade goods at commission meetings every year, they needed information to negotiate and consent to the rules adopted. ICCAT was

Figure 4. Contracting Parties were flagged and prepped to regulate catch during the ICCAT commission meeting in Istanbul, Turkey, in November 2011.

firmly located in the world of diplomacy, a world where delegates attempted to reduce known information "for someone, increase it for someone, or defend someone against it"—without inciting physical violence.[86] To realize the contract between the parties, member states treated the bluefin as the material that secured future relationships, thereby ensuring that the benefits of participating were reciprocated in cash and kind. The experts (re)negotiating these and other contracts on a supranational scale have become "the new diplomats"[87] or, perhaps, the "shadow elite."[88]

Despite a host of supranational regulations issued in almost five decades, the number and size of large pelagic fish and other migratory creatures such as the Kemp's ridley turtle, the oceanic whitetip shark, and the petrel have plummeted under ICCAT's watch. ICCAT's mandate to conserve "tuna and tuna-like fishes" was somewhat of a misnomer. About thirty creatures other than Atlantic bluefin tuna fell under ICCAT's direct concern while I was in the field, among them mackerel, bonito, yellowfin tuna, bigeye tuna, skipjack, albacore, swordfish, and other billfish such as Atlantic blue marlin and Atlantic sailfish (Figure 2). Added recently has been what is called "by-catch" or the various sea turtles, seabirds, and sharks that are caught, killed, or discarded as waste while targeting commercial fish for market.[89] It is remarkable that in the United States alone "by-catch" accounted for an estimated one-third of all creatures caught in the net of industrial fishers, according to a report published in 2005.[90]

Even so, ICCAT did not regulate the catch of all high-seas fish in the Atlantic basin. There were no regulations in place whatsoever for various small tunas during my research.[91] These fish accounted for an estimated 28 percent of the total reported catch in the ICCAT convention area for the thirty-year period between 1980 and 2010. They were a main source of income for artisanal fishers and of food for coastal communities in poor countries.[92] There were no statistical counts rendering how many of them there were, and thus there was no way for the ICCAT modelers to rationalize and forecast the inventory of these fish "stocks."[93]

At the time of my fieldwork, I found that Atlantic bluefin tuna accounted for only 2 percent of all fish caught in ICCAT's convention area but one-third of all regulatory production since the treaty entered into force in 1969. Regulations issued for bluefin tuna did not directly impact all ICCAT members, because only twenty-one Contracting Parties had a quota to legally export this one fish when I was in the field (Appendix B). According to some delegates I met, the attention paid to this creature relative to the other ones in ICCAT's convention area was disproportionate. Member states without a quota for bluefin tuna often considered this discussion immaterial and the obsession in regulating this fish to be self-serving for the rich.

ICCAT made rules that determined catch quotas, or which countries may catch how much of which fish; limited the type of gear used, the vessel size, and the length of the fishing season; harmonized how to transship products at sea and what documents to use for cross-border trade; detailed how often and who may inspect vessels for certain fish while in port or at sea; established the frequency with which a vessel must transmit a signal via satellite to record for compliance purposes where that vessel was at what time. The complex regulations ICCAT issued annually created cumulative demands on staffs from both member states and the secretariat. Since 1996, ICCAT expanded it regulatory scope and gave itself the authority to issue "trade measures"—also known as treaty-based economic sanctions[94]—which prohibited ICCAT member states from importing certain fish from certain countries.[95] The sentiment is: If you want to participate in the global fish trade, you must play by ICCAT rules.

Just as various theorists of international relations claim, ICCAT meetings provided occasions for delegates with diverse interests and priorities to regularly talk, impart views, develop a sense of collegiality, build trust, crack jokes, indulge in shared travel woes, and so on.[96] A common mission (such as "saving" sharks) or profession (such as marine science) could reinforce social bonds. ICCAT meetings provided the physical proximity for representatives of state, industry, and civil society to form potential or tacit coalitions by direct communication, especially when this activity had been made difficult in light of more public controversies.[97] A friendly exchange on the floor of a plenary between China and Japan at the 2012 ICCAT meeting contrasted starkly with the bitter diplomatic row happening publicly over territorial islands in the China Seas. When I asked a prominent EU delegate her thoughts about this "cartographic anxiety," she intoned a single word—with a smile—"Relaaaaxed."[98] International regulatory agreements such as ICCAT helped to smooth over tensions by providing opportunities for cooperation between unlike or even hostile constituencies through regular, predictable, continuous, face-to-face, civil, social encounters. This in itself was a great achievement, even if from time to time these occasions had the potential to damage reputations and strain trust between and within delegations.

The great theorist of bureaucracy Max Weber noted long ago that institutions assert their authority and retain their legal monopoly (over high-seas fish in this case) through regulatory enactment. To create the conditions for ICCAT's own inevitability, Contracting Parties called their authority into being by agreeing to enact the rules to which member states consented during annual commission meetings. This ensured the continuous institutional effectiveness of ICCAT as the international legal authority exercising jurisdiction and managerial power over high-seas fish in the Atlantic basin now and in the future.

This arrangement requires the bluefin to survive. To "save" the bluefin is to "save" not her majesty or the potential for her kin to grow big in size but to secure export markets for the amplification of sea power. Under ICCAT's predatory regime of value, delegates adopted the rationale as a means to control, manipulate, master, and possess precious beings as commodities for the profit of a few at the expense of the livable future of all. Today's reliance on fisheries management as a way to "solve" the overfishing "problem" does more than smooth over the scale of the threat to biodiversity. It denies the limits of what the technocrats can do about stressed biomes to whose vulnerability and depletion they have greatly contributed. *The problem is not which countries have failed to meet their contractual obligations but why the contract fails to value the indispensability of life on the blue planet.* In quite quotidian ways, detailed throughout this book, the practice of valuing nonhuman animals evacuated of life was itself a precondition for subscribing to ICCAT's highly ritualized political order.

Critical Ocean Studies

While the number of institutions governing supranationality has skyrocketed in recent years, situated knowledge about how they work in practice has not. Research on how intergovernmental organizations operate on the ground, from the inside, experienced over time, firsthand, is thin. To paraphrase Sally Falk Moore: to read a treaty is not to understand how the political system works.[99]

My long-term, historically contextualized research takes as its disciplinary base anthropology, while it borrows from and integrates various fields of study, including geography, international relations, the marine sciences, and studies in media, science and technology, and the sociolegal. Without focusing on and blending insights from these knowledge bases, I could not have produced a book occupying the ground from which a genuinely critical account of marine policy making adequate to apprehending the reasons for ecological destruction could be attempted. The intellectual task must be to provide an explanation for how institutional power thrives and continues to operate by rationalizing the domination of vulnerable beings, including sea creatures, at a time of mass extinction. "Critical ocean studies" is the term I believe we should use to characterize this necessarily collective effort.[100] I know what I know mostly through participant observation. My goal was not to "seek representations of daily practice per se . . . but rather the decoding of a historically particular apparatus."[101] In the spirit of multisited ethnography, this book treats ICCAT not as a fixed geographic field site but as a mobile village.[102] I followed ICCAT to Agadir (Morocco), Barcelona (Spain), Brussels (Belgium), Cambridge (UK), Istanbul (Turkey), New York (United States), Paris (France), Tokyo (Japan), Washington, D.C. (United States), Madrid (Spain), and small villages on the coast and in the interior of Spain. Studies embedded in the United Nations system elicit and necessitate multisited ethnography in fieldwork situations where "amorphous and decentralized policy regimes . . . function to regulate" oceanic futures.[103]

But an ethnographically informed account about the regulation of fisheries on the high seas was not only multisited in space. It was compressed in time. At ICCAT, the local and the global collapsed under enormous pressures of time in an institution that cannot be fully apprehended through conventional ethnography alone. When meetings were in session, delegates from across the globe could work for weeks on end, from early morning until late at night, seven days a week. Fieldwork felt like a marathon run in sprints at various times throughout the year on a peripatetic schedule. Immersion was deep but choppy; it was concentrated, finite, fragmented, restricted, not continuous.

Anthropologists have recognized that the binary of the local and the global is unhelpful for research in an interconnected world. Multisited ethnographies explore the ways in which those two poles are joined in practice. Analyses grounded in assumptions about the categorical opposition between biology and culture, nature and society, the human and the nonhuman are rightly being called into question now too. Multispecies ethnography recognizes that knowledge based on the fundamental division between these domains does not accurately capture the realities of the world in which we live.[104] Nonhuman natures cannot possibly be characterized as pristine or untouched when the human imprint is now everywhere on the planet in the Anthropocene. In the multispecies reckoning that is our present, ICCAT was a "contact zone" where the bluefin and the delegate met.[105] I have tried to put multispecies thinking at the center of this study even though—or precisely because—ICCAT's predatory regime of value continually denied nonhuman animals a presence of being. The "human" as the supreme, "exceptional" being on Earth rings hollow when elites in positions of power have demonstrated over decades that they cannot take care of the ocean.

Ethnographic ways of knowing about bureaucracy,[106] the production of international norms,[107] and public policy[108] try, in the words of Nayanika Mathur, "to spell out the difficulties experienced in getting law off the ground in the first place."[109] Armchair academics divorced from experience in the field find it hard to document the microdynamics of power that inform how technocratic elites develop comprehensive, rule-bound norms in supranational regimes whose bureaucratic processes sow the seeds of their own contradiction.[110] The ways the timeless concepts of "national interest" structured the work of state delegations, for example, were, at ICCAT, far from straightforwardly evident or agnostic in their effects.

To understand the inner workings of the world's most powerful fisheries management organization, research also required moves beyond the ethnographic. An investigation of the "frames" circulating in the news media clarified the public pressure ICCAT was under from mainstream environmentalists who organized transnational campaigns to "save" the bluefin. (See chapter 3.) Archival records available online and at the ICCAT secretariat assisted in piecing together ICCAT's early history. (See chapters 1 and 2.) These research components were crucial in helping me to connect the dots widely separated in time and space. These dots, discerned ethnographically, once connected, allowed me to assemble as best I could a complex configuration taking the form of a bureaucratic institution composed of member states dedicated in name to marine conservation.[111]

ICCAT's regime of value is entangled in a world well beyond it, even as it gives expression to the world in which we live. I try to offer a worm's-eye view of the intense prospecting for "red gold": the ways deals were struck, reputations made, careers advanced, institutions produced, nations built, economies developed, empires emboldened, and reciprocity returned in kind through the material force of a globally regulated, high-priced fish. Over the years I heard stories of rich ICCAT member states throwing their weight in military maneuvers at sea to intimidate artisanal fishers. I heard stories of other bully tactics, such as bullets sent by post to environmentalists and white chrysanthemums placed at a marine advocate's desk while on break during an ICCAT commission meeting, as if to foreshadow an impending funeral. I heard stories of fish traders also trafficking arms, drugs, and migrants. I saw intelligence reports about ICCAT delegates issued by government spy agencies.

I have no way of verifying some of these claims, and instead treat them as part of the ICCAT mythology: not to suggest that they were untrue but that they informed the stories the delegates told to make sense of a diplomatic world of smoke and mirrors. I understand these myths not as limits to my knowledge claims but as windows into the way the delegates experienced ocean governance behind closed doors. There were "trade secrets" to keep, I was told. ICCAT delegates wondered aloud, but in a whisper: Who works for whom? Has she been paid off? What led to the collapse of the relationship between financier X and "tuna farmer" Y? As the institutional and spatial journeys of commodities grow more complex through globalization—as the mutual alienation among producers, traders, and consumers increases—the more frequently I experienced the "culturally informed mythologies about commodity flow." Commonplace across delegations, mythologies often acquired a particularly intense valence "when the spatial, cognitive, or institutional distances between production, distribution, and consumption [were] great."[112] The farther apart people were from each other in the trade of bluefin tuna, the more apt they were to imagine vivid and conspiratorial worlds about ICCAT and each other.

The challenge of gaining access to international organizations as a researcher is real, which explains, in part, why ethnographic accounts of global governance are few.[113] ICCAT forbids journalists and members of the public from attending its meetings. In June 2010, I began triangulating among various delegations in search of access, and, like a football, was passed and punted between delegations of state and civil society as I sought admission to the rooms where I could watch how supranational marine policy was debated, brokered, and decided. I rode the merry-go-round for three months until September 2010 when a member of the U.S. delegation suggested by e-mail that I apply for "observer status" as a delegate representing an institute at my university. I did. My affiliation as a fellow at the Institute for Public Knowledge (IPK) at New York University got me in. The ICCAT commission accredited IPK as an observer in October 2010.[114] Unlike other observers, IPK as a delegation from a U.S.-based research institute never had an official policy or a position dictating how I would act onstage or offstage. I never made an intervention on the floor of an ICCAT meeting, nor did I circulate documents of any kind.

I conducted research as a nonaligned observer, and this indelibly shaped my findings. My credibility before other delegates was derived in part from being an independent observer with no state, industry, or "green" affiliation. Even so, my status foreclosed important possibilities for exploration. I did not have access, for instance, to the special conference rooms organized by rich countries ("delegation rooms") or to password-protected Web sites available only to delegates from ICCAT member states. I learned in the very first days of observing ICCAT's decision-making process that ICCAT delegates also observed me. "I see you were chatting with X [industry representative from Spain]," said one member of the U.S. delegation at a commission meeting in November 2010. "What are you typing all the time?" asked another delegate from the EU while I followed the proceedings of an ICCAT commission meeting in November 2012. Sometimes I felt like a spy more than a rightful observer of what global elites were doing about an ocean emptying of big fish.[115] My experience was as transient and limited as finance capital, and as stateless and studied as the bluefin.

Because of lack of access, the major player hovering over this book rarely comes into focus. The multinational corporation Mitsubishi is said to control 40 percent of the bluefin tuna trade. My encounter with one of its representatives came only once by happenstance while at a dinner party during a research trip to Tokyo. Mitsubishi has the ear of the Japanese delegation, just as it does the delegations of Spain and Morocco, among others. Finance capital knows no boundaries. It is so dispersed in the ICCAT apparatus that it is empirically difficult—and misguided—to treat nation-states as sealed units.[116] National governments—whether rich or poor—respond to the demands of business.[117] This fact is not itself a scandal. It only becomes scandalous when states work in favor of industry in ways that are disproportionately harmful to others.[118]

Over the course of three years, beginning in fall 2010, I participated in and observed fourteen ICCAT and ICCAT-related meetings on four continents. The expense of travel combined with the frequency of meetings limited how many meetings I could attend. I also conducted thirty-eight semistructured interviews with representatives from state, industry, and civil society outside of meeting events.[119] Only four of these interviews were completely on the record. The last of these interviews took place in summer 2015.

The vast majority of delegates shared with me their thoughts informally. They were careful to separate an "official" narrative from their "personal" viewpoint.[120] This is one of the reasons why I disguise attributions throughout the book. I rarely name anyone directly, and in most spots only reference general dates to protect anonymity. Within these constraints, I have attempted a microscaled account of ocean governance that sheds light on patterns in institutional frames[121] and in the "collective agency"[122] that binds ICCAT delegates interested in realizing the rewards and value structures of commodity empires. Nonanthropologists, concerned about the reproducibility of these findings, will have to be content to find it in how reasonably recognizable my descriptions of ICCAT are, compared to their own or to those of other researchers.[123]

The density of social networks cultivated over time was itself a source of value enabling delegates to exchange information, define policy in the making, and ascertain positions. As a participant in, and observer of, the ICCAT world, I too formed my own network whose value informs this book. The relationships I developed in the field shaped what questions I could ask of delegates and what happenings I could observe. I had to collaborate with and respect the views of all ICCAT delegates, even the ones who offended my sensibilities. Some policy makers I met thought that the climate crisis was a hoax and others that the ocean was teeming with fish. It may surprise the reader to know that some of my most productive conversations were with industry representatives. They and many other constituencies wanted ICCAT to succeed—as do I. This book tells my own truth, one that is situated within and inherently influenced by a subject position I cannot ever fully escape: a woman, a U.S. citizen, an academic, white by today's standards in America, of relative privilege, still able-bodied.[124]

"Ahhh, so you're a fisherman's daughter," an environmentalist once said to me, as if this information explained me to him. I grew up in a family of recreational fishers. My grandfather taught me how to bait hooks. The blowfish that used to balloon and make us giggle when I was a child are now gone from Shinnecock Bay on eastern Long Island. I am concerned about a world without "the very elder gods" and the other fish—big and small, mega and mini—and the flora on which all beings mutually depend. I believe this background makes me not "biased" but informed, curious, and committed to sea worlds.

While I remain committed to a plausible, sensitive account of IC-CAT based primarily in prolonged first-person participant observation, I do not want to rationalize the violence of managed extinction or become allied with, or falsely neutral about, the delegates who passively abide this stance in a commodified regime of value. I am not a clerk recording the minutiae of ICCAT events, and yet I am not innocent. The conditions that enabled me to observe and participate in ICCAT goings-on already made me complicit in commodity empires. As a New York–based academic trained in the United States, I too am part of the global elite. After years of contemplation, I find it more productive "not to detour around, but to move through, and hopefully beyond" ICCAT's predatory regime of value. The aim is not to grab hold of and overpower the alienation but to evoke it and make it more fully present, not to ignore or disavow it but to let it out and be.[125] If the reader is left with feelings of despair or frustration, then I know the book has done its job. I know it did the work of unsettling comfortable assumptions about the viability of technoscientific solutions to overfishing when attuned to an exterminatory regime of value.[126]

The ICCAT "Game" Play by Play

As do Vladimir and Estragon in Samuel Beckett's play *Waiting for Godot*, the characters that inhabit this book represent the effort to hold a terrible silence at bay by keeping themselves very busy. ICCAT delegates waited for the mystery of the ocean to reveal itself through faith in what fisheries science could deliver. Meanwhile, the ICCAT commission worked very hard to referee economic growth by developing marine policy for nation-states big and small, rich and poor, in the name of marine conservation. That the ocean has been systematically emptied of big fish without sustained public outcry signals how alienated the modern condition has become from life considered part of Nature.[127]

This book speaks to a particular moment in ICCAT's history. It cannot comment on more recent institutional developments but can be, I hope, a useful guide for them.[128] This is a story not of fishing or of fishers, but of their regulation.[129] I realize that the attention I pay to Atlantic bluefin tuna runs the risk of singling out one fish to speak for what is happening to them all.[130] I do not want to reproduce the logic privileging the survival of a narrow range of fish consumed by elites while the fate of the other ones, especially those on which the poor depend, remains unclear. I want instead to capture the contours by which fisheries management operates—isolating profitable commercial fish one by one, species by species—by refusing to account for and value the crucial ways nonhuman animal beings interact at any given time in the very biome we share.[131] This book should not be read as a narrow, all-knowing critique of ICCAT. Rather, it should be read metonymically, as lesson(s) for the battles ahead.

Chapter 1, "A History of the Bluefin Tuna Trade," argues, in the *longue durée*, that ICCAT must be located in the complementary projects of capitalism *and* global empire. It charts the emergence of ICCAT's regime of value over great expanses of time, and shows how, in the long march of Anglo-European history, alienation from nonhuman animal beings helped to organize the dominant mode of fisheries management today. Entangled histories provide the necessary backdrop for understanding why concern over an already stressed bluefin tuna instigated ICCAT's formation from day one. By the 1970s, just when the sushi economy went global, ICCAT was poised to legally permit "animal capital" to systematically accumulate for commodity empires. I conclude by suggesting that ICCAT member states have become not only market *managers* supplying high-seas fish. They have also become market *speculators*—that is, attuned to exploiting the risks associated with financing the trade of underlying assets in a futures market dedicated to fish "stocks" in the open ocean.

Chapter 2, "A 'Stock' Splits," illuminates how ICCAT legalized a predatory regime of value by distributing the fruits of extractive capitalism according to power structures from older eras. Since the early 1980s, at the time of the adoption of the revised United Nations Convention on the Law of the Sea (UNCLOS III), ICCAT member states agreed to split the bluefin tuna "stock" in two: one set of quotas for the western Atlantic and, later, another for the eastern Atlantic. Export quotas for Atlantic bluefin tuna—allocated most to rich countries—regularized the global market in "red gold" by treating the bluefin as the national property of the member states signing ICCAT's trade agreement. It shows that biopower resides not only in legal ends or doctrine, but in means or process. Hierarchies of power embedded in regulatory processes shape regulatory outcomes, and vice versa. The tragedy of lost generations of big fish, I conclude, is rooted not in the commons but in the commodity form.

Chapter 3, "Saving the Glamour Fish," asks: how did ICCAT reproduce a predatory regime of value, even when confronted by its most trenchant critics in the public sphere? It re-creates ICCAT's more recent history by recalling the environmental campaigns endeavoring, in their words, to "save" Atlantic bluefin tuna from the clutch of "commercial extinction." Since the early 1990s, a social drama recurred in such news outlets as the *New York Times*. I call this drama "the savior plot." Bourgeois readers came to know ICCAT as a villain preying on the bluefin, as if an innocent victim in Nature. Environmentalists appeared as the ones who could rescue her from the existential horror of annihilation. Simplified frames—about which ICCAT delegates were fully aware—spoke of overfishing as a management problem the ICCAT commission could reasonably solve on its own. More trenchant, systemic critiques rooted in environmental justice were outside the reach of conventional frames in mainstream media accounts.

Chapter 4, "Alibis for Extermination," shows how ICCAT performed and justified its legal mandate to regulate the catch of high-seas commercial fish by invoking what appeared to its constituents as the independent, neutral authority of fisheries science. This chapter ethnographically describes how ICCAT's predatory regime of value was never challenged in a scientific advisory committee. In that committee, statisticians instead retrofitted data to probabilistic models like actuarialists to inventory "red gold" and to assess the risk of catch during a bluefin tuna "stock assessment" in 2012. ICCAT's deep investment in fisheries science gave member states an alibi for the rapid loss of big fish on the high seas.

Chapter 5, "The Libyan Caper," recounts an important moment in ICCAT's history when rich and rogue delegations played the ICCAT "game" by toying with rules of procedure, most notably consensus, during a crucial commission meeting in 2010. Ethnography substantiates what Arjun Appadurai calls a "tournament of value."[132] These periodic moments of intense sociality demonstrate how those of privilege and power expressed status rivalries. At issue in the ICCAT "game" was not only the character of the bluefin as the central token of value in the sushi economy. It was also the status, rank, reputation, and renown of organizations, of entire delegations, and even of individual delegates.

Although the bluefin is the real protagonist of this book, she is forced to appear in it almost always as a phantom. Her majesty vanishes from these pages. ICCAT delegates must learn to deny and erase her awesomeness so that they may do the work required of an international regulatory agreement mandated to referee the global fish trade. Can ICCAT be remade and change from within, or do its regulatory practices short-circuit any possibility for meaningful transformation? My Conclusion, "All Hands on Deck," stages the following thought experiment: it asks the reader to contemplate the prospect of the last remaining bluefin tuna in the ocean. This bluefin is alone with only jellyfish and worms; separated from her school; swimming aimlessly in warmer, toxic, acidic waters; ingesting (micro)plastic; frantic to find what forage fish remain; unable to meet a mate. This figure in the mind is meant not as a cheap scare tactic but as a way to forge solidarity and incite possibilities. In the words of Haraway, it asks readers "to think together anew across differences of historical position and of kinds of knowledge and expertise."[133] The Conclusion attempts a realistic haunting to figure the deepest contradiction now spirited over by the work of ICCAT's exterminatory regime of value. I hope readers will open themselves up to reckoning with their own place in predation and question for themselves the false promises of a technocratic escape being offered by the current institutions entrusted with the profitable conservation of life on Earth. I hope readers will arrive at a place where they see that the problem is not the institution per se but the values that arrange it.

The owl of Minerva has spread her wings. I invite the reader to reflect on her silent flight in this time of extinction. In the end, as high modernizers cling to bureaucratic loyalties and universalize their faith in "progress," realized through particular applications of law and fisheries science, they and most of the rest of us remain bound to an international state system financed by the "deep establishment."[134] But in actuality, the costs of maintaining the profitability of "red gold" are too tremendous for the present systems of accounting. In the midst of this attempt to catch a glimpse of ocean governance, the reader may entertain the stray thought that the dominant systems of value are broken. Therein lies my commitment to writing this book.

1

~~~

A History *of the* Bluefin Tuna Trade

THE EMERGENCE *of* COMMODITY EMPIRES

At the Secretariat

The world's most powerful regional fisheries management organization (RFMO)—the International Commission for the Conservation of Atlantic Tunas (ICCAT)—finds its terrestrial home on the sixth and seventh floors of a building it shares with the Spanish Institute of Oceanography in the landlocked capital of Madrid. Even the secretariat's physical plant speaks to ICCAT's tortured presence in the cosmopolitan world of ocean governance. The imaginative designs of its neighbors—a five-star luxury hotel and one of the capital's architectural marvels—flank the secretariat on either side and render it unnoticeable. To the east, vertical stripes in a candied tart of red, orange, yellow, and purple stretch the twelve stories of the Hotel Silken, overshadowing the secretariat's lackluster, squat brick building. Juxtaposed with the hotel's rainbowed exterior to the west is the cold apartment complex in gray concrete known as Torres Blancas. The architect Francisco Javier Sáenz de Oiza conceived of it in the styles of rationalism and organicism in the 1960s for residents who could appear in the space-age animated sitcom *The Jetsons*. To these environs, delegates of ICCAT member states have come to speculate in the future of fish "stocks."

Similar to other supranational regulatory regimes, the ICCAT secretariat performs administrative functions to implement rules for the commission, but it is not directly subject to the control of the states that are party to the treaty.[1] It oversees and directs ICCAT's organizational routine. It plans all meetings of various auxiliary bodies, setting deadlines and providing opportunities for discussion; it prepares meeting agendas; it documents and publishes meeting results; and it collects and oversees an enormous amount of information related to scientific research and to fishing activity, among other activities.[2] It records its work in a mass of bureaucratic paperwork, some of which is available on the ICCAT Web site for publics that understand techno-legalese.

The executive secretary oversees the secretariat's day-to-day operations. The journalist Sasha Issenberg writes that he worked "out of an office smartly appointed with mid-century modern furniture and rich woods."[3] These extravagances were absent elsewhere at the secretariat. Headquarters felt more like a throwback to an underfunded marine biology department at a university from the 1960s than an intergovernmental organization mandated to regulate the catch of profitable commercial fish on the high seas, including "stocks" of "red gold." My first impressions were of discounted furniture, steel bookcases built to last, artificial light from fluorescent bulbs, speckled linoleum floors, bent Venetian blinds clanking in the breeze, old air-conditioning units moaning while suspended in thinly paned windows, taxonomies of different kinds of fish posted throughout the halls, some faded by age and intense sunlight. Little pedestrian traffic flowed past the building, but one of the main arteries leading into the Spanish capital ends where ICCAT begins. Passersby had no idea that north of the highway the secretariat was a center for the administration of one of the world's great commodity empires.

Commodity Empires in the *Longue Durée*

What can "big histories" tell us about ICCAT?[4] What role did Atlantic bluefin tuna play in instigating the emergence of an intergovernmental organization mandated to regulate the global supply of high-seas fish? Happenings in centuries long past are remarkably helpful for understanding why ICCAT regulates in the present the catch of a fish now worth its weight in gold. Here, over a long period, elements of the power to render a living being capital have dovetailed with the administrative history of imperial states in the long march of Anglo-European history. From this slow-burning kindling came the global firestorm of coordinated wealth production I believe is helpful to recognize today as "commodity empires."[5]

The next two chapters chronicle the piecemeal historical process by which elites have constructed a juridical apparatus known as ICCAT. Taken together, these chapters emphasize "the importance of historical contingency in human affairs." They allow later chapters, ethnographically grounded, to illuminate "how moments of rupture, conflict and discord result in power inequalities concealed through different political technologies."[6] These technologies include bureaucratic paperwork, fisheries science (chapter 4), and rules of procedure (chapter 5). ICCAT is the product of circumstances inherited over great expanses of time. Contingent developments in the *longue durée* have contributed to the formation of an administrative infrastructure that contemporary finance uses to make global wealth accumulation so profoundly successful.

In this chapter, I show, following Amitav Ghosh, that the Anthropocene as an organizing paradigm must be situated within the histories of both capitalist development and global empire. That is, the Anthropocene must be seen as thoroughly permeated by the way power over capital is distributed. It must be understood as based on "an aspiration to dominance on the part of some of the most important structures of the world's most powerful states."[7] This approach allows for an account of capitalist development, contradiction, and crisis situated squarely within the organizational imperative for a zone where elites may oversee and coordinate the global commodities trade. Here ICCAT member states—in the growing network of ocean governance—constituted markets, legalized the rules of trade, and formalized imperial orders by regulating and distributing wealth and power derived from fishing on the high seas. This is the proper introduction to ICCAT, a supranational regulatory agency about which most citizens know nothing, despite the fact that its member states have accelerated at an exterminatory rate the slaughter of high-seas fish since the regime's founding in the 1960s. Since then, the ICCAT apparatus has conditioned which fish people eat and how much people pay for them.

This chapter charts the *longue durée* of the emergence of an exterminatory regime of value that, over time, has come to dominate how marine conservation is conceived in ICCAT's regulatory zone. It narrates, unevenly, material developments from centuries past, taking the reader through the 1970s. This history will show that the Atlantic bluefin tuna has long been an object of trade and commerce in the Mediterranean basin. It was anxiety about the decline in this one fish that galvanized ICCAT's formation in the years following the end of World War II. The reader cannot appreciate why ICCAT formed when it did without an account of trade in bluefin tuna over millennia. The Atlantic bluefin tuna fishery was thoroughly entrenched and deeply braided with the social lives and cultures of peoples in the Mediterranean basin. This history must include the foundational moment in the scientific management of commercial fish. In the 1950s, nation-states under the influence of the United States agreed to adopt as a "harvesting" limit the juridical formula of "maximum sustainable yield" (MSY) under the United Nations Convention on the Law of the Sea (UNCLOS I). This set the stage to legally constitute the export markets of ICCAT member states during the Great Acceleration. Signatories to the ICCAT convention were poised to legally permit the growth of national economies sourced from the ocean with the rise of the global sushi economy during the 1970s. From its very beginning, ICCAT was flexible enough to meet the demands of member states as they negotiated changes in political economy, including the major shift in global wealth production represented by the recent ascendancy of finance capitalism.

The Ocean Does Not Exist outside History

I want to begin this journey through geological time not on land but at sea.[8] I do this not to develop an origin myth or tale of creation but to counter "the enduring assumption that the ocean exists outside of history." W. Jeffrey Bolster writes: "Relegated to the role of sublime scene or means of conveyance, and shorn of its genuine mysteries and capacity for change, the ocean appears in most histories as a two-dimensional air–sea interface—a zone for vessel operations and a means of cultural interactions."[9] Storytelling, he suggests, must avoid the false binaries of nature and culture, animal and human, as if historical change can be traced *either* to pressures wrought by people (overfishing, pollution) *or* to natural environmental shifts (the number of fish in the sea fluctuates). Rather than treat these domains as separate and watertight, they are better understood as fluid, entangled, interacting.[10] What the bluefin does and what happens to her when the dominant society treats her in a certain way implies that she carries on, undergoing constant modulation, changing in her environment—just as she changes people in theirs.

The paleontological record indicates that in the Early Tertiary Period some sixty-five to two million years ago, when mammals rose to occupy Earth after the extinction of the dinosaurs, the giant bluefin tuna came into being. In the Cenozoic Era or "Age of Recent Life," of which the Tertiary is part, the physical characteristics of the ocean as we know it first appeared: its gyres, its thermal structure, its ridges and folds in the underwater world. Then the bluefin began to explore the open ocean and dwell in the newly formed Atlantic Ocean, which opened from the very old Tethys Sea, once home to various creatures during the Mesozoic Era or "Age of Medieval Life." As she parted ways with her distant cousins, the bonitos, radiating outward on vast migrations while still visiting coasts from time to time, under the influence of changing paleo-ecological and paleoceanographic conditions, the bluefin developed what characterizes in the present all tuna kin: the capacity to generate and conserve heat. Her muscles are not chilled in cold water. This was itself an evolutionary response to the cooling water temperatures of the Tertiary Period.[11]

The bluefin over time refined her heating system. She is endothermic, like you and me, capable of generating internal heat. She has learned to elevate parts of her body temperature higher than surrounding water through countercurrent heat exchangers. She can rearrange the way her veins and arteries course through her body. Her muscles spring into action but recover oxygen quickly. They allow her to swim like a bullet, from the tropics to polar latitudes, like a hydrofoil thrusting in seawater, like an Olympian runner bursting into sprints without ever the thought of a finish line.[12] She survives in perpetual motion. She must swim and move forward, always, to stay alive, lest she suffocate and sink to the ocean floor.

To endure a world in constant motion, swimming lightning fast, the bluefin developed an exceptionally large heart. Her heart runs the ambient temperature of surrounding water as it receives oxygen from gills, which all fish hearts do. Yet the bluefin can keep her heart beating even when encountering frigid temperatures in the deep ocean, which in people would cause a heart attack. It is precisely this ability that allows her to inspect all corners of the ocean.[13] Her cousin, the bigeye tuna, is the only one that can match her vertical range. Up and down, left and right, far, very far, surviving pressures when on a downward dive of more than three thousand feet (one thousand meters), two and a half times the height of the Empire State Building, all grown up, she chases bony fish as she gets older, metabolizing them quickly, no longer satisfied by zooplankton as when she was young. She finds companions of similar size who travel with her in packs. Perhaps it is among this group with whom she finds a mate and spawns at night in warm waters certain times of year.

The bluefin makes one of the most extensive migrations of any fish on the planet. The largest of all tuna, capable of weighing well over a ton, the Atlantic bluefin today traverses the ocean from the Gulf of Mexico to Newfoundland, from the Canary Islands of Spain off continental Africa to Iceland's southern waters, from the Mediterranean Sea to Norway. One bluefin tuna tagged off the northwestern Bahamas in June 1969 journeyed at least 6,600 nautical miles (7,612 miles or 12,250 kilometers) to Argentina in February 1973—one of the longest migrations ever recorded.[14] She has been known to voyage from the West to the East Atlantic in only forty days.[15] Amid her epic travels, perhaps she returns to a place in the sea she calls home, where and how often we do not know with certainty. I wonder if she sleeps, who she plays with, how she expresses her sociality, what her response to people is.

The bluefin is not a fish among fish. She is something else. Our forebears knew as much.

Notes on Commodity Empires

That empires have been made and secured through fish is a phenomenon of the historical record as much as it is of the ethnographic present.[16] Over the centuries, the Atlantic bluefin tuna has built empires. The Roman, Moorish, Portuguese, Spanish, and French empires, for example, are all indebted to the bluefin in ways still felt in ICCAT meeting rooms. Empire effectively characterizes ICCAT's social life and regulatory culture in ways most delegates today would readily recognize even though they may not use this language.

Empire is not only an idea bandied about as academic fashion. It is a practice of the global historical imagination that today implicates states as much as it defines nations. Empire not only characterizes the actions of the powerful in which the usual suspects of the former metropole continue to strong-arm the weak of the periphery using the ICCAT apparatus. Aspirations to commodity empires in the present help explain delegates' deep investment in and commitment to the ICCAT regime. All delegates come to meetings with their own particular imperial histories—American, Ashanti, Chinese, Dutch, Egyptian, French, Greek, Japanese, Ottoman, Persian, Portuguese, Roman, Spanish, and so on—no matter how delicate and fleeting, uncertain and enigmatic, uneven and episodic many of them have been through the years.

I found while in the field that the bluefin as commodity enlivened and stoked the pride, the legacy, the memory, the nostalgia, the aspiration, the apology, the resentment, and, not least, the nationalism underlying "resource" extraction. The promise of return to some kind of glory or the rebuke, containment, and resistance to that desire characterized each ICCAT commission meeting I attended. Perhaps this year the ICCAT commission will affirm, redeem, recalibrate, or recognize as true equals a delegation's rightful place at the negotiating table. By performing this ritual at commission meetings year after year (discussed in chapter 5), ICCAT called empire(s) into being in the service of the nation through monetized life-forms. This occurred even as finance capital crosscut and left the nation not as whole as imperial discourse insists on imagining in its self-representations.

By bringing together political economy, law, and material culture, this chapter shows that ICCAT member states were concerned not with "the empire of things" but with empire *through* things—*which emphasizes the process for empires' achievement*.[17] Seen in this light, the bluefin has become a material force, a "vector of law," mediating aspirations for commodity empires among ICCAT member states.[18] The point is not to reduce the ICCAT regime to mere economy—as if it was all about monied interest—but to underscore the cultural norms, commitments, and status plays making regulatory oversight of economic action possible.

After all, the vast majority of ICCAT delegates—even from rich trading areas—did not themselves make fortunes from "red gold."

"The world order is expressed as a juridical formation," write Michael Hardt and Antonio Negri.[19] ICCAT is one of these formations. Ecological anxieties felt across the globe demand diplomatic action. But as the great theorist of nationalism Benedict Anderson has long argued, the nation is a "limit" concept that has "finite, if elastic, boundaries, beyond which lie other nations. *No nation imagines itself as coterminous with [human]kind.*" The Westphalian state system—which depends on the cultural production of the nation as a people or community so that it may declare itself sovereign upon the recognition of other states—is a peculiar basis upon which to broker deals on behalf of planetary life-forms. The deep, horizontal comradeship imagined as nation may have made possible feel-good moments within ICCAT delegations.[20] But nationalism also reinforced and sharpened the divisions between member states as each sought to legalize its claim to a share of whatever "resources" remained to the exclusion of others. Diplomats on the world stage of environmental governance are now stuck in a juridical framework that vies for control of the economy based on *thinking the national, even the global, without ever thinking the planet.*

Commodity empires signal an important element of ICCAT's milieu beholden to various histories and their accompanying spatial imaginaries. Controlling oceanic space for the movement of goods has long been a practice of empire building during capitalist expansion, whether those "resources" were slaves or sugar, cotton or cod centuries earlier.[21] Who lords over the ocean and controls its fish, whales, oils, gases, shipping lanes, and undersea cables is a way that imperial power struggles have long played themselves out by nonterritorial means.

In a too-little-known essay first published in 1942, *Land and Sea*, Carl Schmitt discusses the centrality of the sea in the advent of global empires. He draws on the work of U.S. Admiral Alfred Thayer Mahan, who claimed in the nineteenth century that a strong navy made a strong state.[22] More recently, Elizabeth Mancke argues through the European example that "control of oceanic space has become not just a commercial question but part of the construction of power in the European state system."[23] Of course, the Anglo-Europeans did not invent this phenomenon. Imperial powers have long used fishing boats to assert their might: the Japanese control of German islands post–World War I and more recent efforts by the Chinese to claim as their own the entire South China Sea illustrate the point. To cite Sir Walter Raleigh's famous dictum from the seventeenth century: "For whosoever commands the sea commands the trade; whosoever commands the trade of the world commands the riches of the world, and consequently the world itself."

In the long run, we need to know how common publics, state bureaucrats, marine scientists, industrial fishers, environmentalists, and a host of other actors complicit in ICCAT's regulatory action are drawn into—and consent to—the way fisheries management organized by the nation-state has administered extermination over time for commodity empires. We must have an account of why people acquiesce to ICCAT's exterminatory regime of value. Concepts of culture—the imagined nation, the excitable empire—do this kind of work.

While histories help to organize the present-day circulation of Atlantic bluefin tuna in the global market, commodity empires are not a timeless universal. History does not repeat itself. There is something distinctly new about the moment in which commodity empires operate today. Commodity empires must now be understood through the conceptual framework of *geoeconomics*, a theme emphasized throughout this book. I ally myself with what Deborah Cowen and Neil Smith describe as a "conception of space, power and security, which sees geopolitical forms recalibrated by market logics," producing new forms of sociality in the process. These authors recast, rather than replace, geopolitical calculation within the projects of global trade and national security. Geopolitics is not lost at all here.[24] Instead, it is better specified. In the realm of commodity empires today, when access to global trade in "natural-resource" markets is limited, when food and border security are the concern of the nation-state, when capital is no longer territorially based, new forms of governance and new attendant forms of sociality have emerged in zones such as ICCAT.[25]

Geoeconomics as a way to conceptualize the contemporary moment becomes all the more convincing when considering work by Timothy Mitchell. He argues that "the economy" as a category of representation is a recent historical phenomenon. The term *economy* is new to political discourse, so much so that it did not enter political lexicons, let alone political debates, until the 1950s. He writes: "The concept of the economy . . . [as] the structure or totality of relations of production, distribution and consumption of goods and services within a given country or region . . . only dates from the mid twentieth century."[26] Not until the devastation wrought by World War II—itself a crisis of overproduction and stagnation from the 1930s—did the economy as we know it take shape. Similarly, as Arturo Escobar writes, by the early 1950s, "a growing will to transform drastically two-thirds of the world in the pursuit of the goal of material prosperity and economic progress . . . had become hegemonic [in] circles of power" at the United Nations.[27] That the economy emerged as a preoccupation of the nation-state at the same time as the Anthropocene's Great Acceleration should not be lost on the reader.

According to Mitchell's probing analysis, state officials assumed that mathematical experts could ease anxiety about wealth by predicting change in a newfound economy according to number. A system of aggregates mapped onto the geographic space of the nation-state. Such figures as the GDP of India, Mexico, and France, for example, could represent the size, structure, and growth of economy as a scalable abstraction ready for intervention. It was as if economic discourse produced real stoppers, levers, and pivots in a machine to be managed by technicians. A politics by economy, Mitchell argues, "provided a new way for the nation-state to represent itself," allowed for a new conception of international order, and instilled a new priority in a politics about growth.[28] The United Nations, the World Bank, the International Monetary Fund—even a relatively unknown institution such as ICCAT—formed in the image of separate nation-states marked by distinct, measurable economies, some with imperial ambitions, taking as their aim growth though such market mechanisms as the global trade of commodified fish.[29]

The aspiration for commodity empires is not a mere metaphor for power. Rather, when elevated to the level of the analytic, it constructs a host of practices and ways of being in the world hammered out in zones such as ICCAT.[30] Commodity empires engineered geoeconomically gesture to a form of political organization encompassing multiple polities at a time when capital markets are no longer contained within the territory of any one of them. Hardt and Negri have famously called this formation "Empire." A *commodity* empire resembles this understanding: it "establishes no territorial center of power and does not rely on fixed boundaries or barriers. It is a *decentered* and *deterritorializing* apparatus of rule."[31] As member states cope with the risk associated with supplying a futures market in high-seas fish, ICCAT asserts its imperial sway too, in part because member states accept what they otherwise would not in order to maintain their good standing in the community of nations.

In commodity empires today, power in the main is exercised, absorbed, (re)articulated, interpreted, dispersed *biopolitically*. Biopower means, writes Michel Foucault, "power over life," or the forms of power exercised over living beings: some bodies cared for, their lives maximized; other bodies discounted, their lives considered threats to society (read: the black, the brown, the disabled, the feminized, the transgendered). Biopolitics is concerned with people as members of an abstract "population" in which biological and presumed pathological features link up to national policy in the exercise of power.[32] In this formulation, when situated historically, the state is just an event or episode in government, a kind of rationality with an arsenal of corresponding practices that, since the eighteenth century, has intervened at the level of the "population," rather than at the level of the individual.[33] Although Foucault did not consider the nonhuman animal central to his biopolitical framework, we might. The violence wrought by the organized extermination of the nonhuman animal suggests that the "*practice* of Empire is continually bathed in blood, [but] the *concept* of Empire is always dedicated to peace—a perpetual and universal peace outside of history."[34] The task remaining in this chapter is to write a history of this paradox, of violent peace, by treating the bluefin as a source of wealth and target of power, abstracted by her "population," which has led to the managed extinction of big fish. All this shares an affinity with the theoretical armature of biopower.[35]

Commodity Empires Permit "Animal Capital" to Accumulate

Some of the ICCAT delegates I met recognized what the archives from the Mediterranean attest, namely, that Atlantic bluefin tuna has been caught in webs of exchange for millennia.[36] The chronological depth, density, and complexity of the Atlantic bluefin tuna trade indicate that bluefin tuna is one of the earliest recorded fisheries known and regulated since the time of classical antiquity. History—not geography—unfolding over vast expanses of time structured the design of ICCAT as an apparatus for the accumulation of "animal capital."[37] This holds true despite recent attempts to erase the deep past of bluefin tuna fishing for political expediency. (See chapter 2.)

"Big histories are always best told through insistent, if humble, details," Anna Tsing writes.[38] Indeed, artifacts as far back as the seventeenth century BCE and ceramics from the twelfth century BCE depict the bluefin's capture.[39] Coins stamped with her image signified her importance to the Phoenicians and their heirs, the Carthaginians.[40] The full range of her symbolic, mythical, and religious connotations remains unknown.[41] Her image still resonates with ancient meanings in places such as Kyzikos in present-day Turkey and Gela in southern Sicily.[42]

That currency is marked at all with the image of a tuna is not a forgotten relic or a curiosity from the archaeological record at ICCAT. I heard from more than one ICCAT delegate that a member of Croatia's delegation (before joining the European Union) waved in the mid-2000s one of these coins on the floor of a commission meeting to justify his share of "red gold," stating: "We love our tuna more than our women." The gendered dynamics of "resource" extraction exhibited by this and other ICCAT delegates illustrates why the masculinist ambition to control and dominate feminized natures is key to understanding the Anthropocene, now more than ever amid the velocities of the Great Acceleration.[43]

Remarkably, Aristotle alluded to the tuna trade more than two millennia ago. I quote a passage attributed to him dated roughly 350 BCE:

> The Phoenicians who colonized the place called Gadeira [present-day Cadiz, Spain] arrived at certain deserted regions . . . [where] they find an exceptional number of tunas, incredible both in size and fatness, whenever they run ashore. They pickle these, store them in casks, and bring them back to Carthage. These are the only tuna which the Carthaginians do not export; they consume them themselves because of their excellent taste.[44]

In Greek city-states, evidence of trade and dedicated markets indicated that duties were assessed on tuna in the mid-third century BCE. Tax and other regulations on the sale of seafood were important sources of public revenue.[45] The Greek geographer Strabo describes how the wealth of the ancient city of Byzantium came from tuna.[46] The "good fortune" of Corcyra in what is present-day Corfu also rested on investment in tuna fish.[47]

For the ancient Romans, Pliny the Elder wrote in the first century CE of tunas' coastal migration routes,[48] chronicling how schools of tuna impeded the navigation of fleets commanded by Alexander the Great.[49] His contemporary Plutarch described how they ran in packs "to stay together and for love of one another."[50] Fascination with bluefin tuna extended to emperors and nobility in the Roman Empire. They spent enormous sums of money on large fish, which they adorned with jewels and placed on platters carried into banquets to the accompaniment of flute and pipe. The archaeologist Brian Fagan writes: "The larger the fish, the more valued it was; the largest were reserved for emperors and kings."[51] And the largest tuna of all, as we know, would have been the Atlantic bluefin.

It is remarkable that even millennia ago the sale of and record keeping for bluefin tuna were systematic, "the result of organized collective action." Already by the fifth century BCE the bluefin was an "asset capable of being turned into disposable wealth." Peregrine Horden and Nicholas Purcell state: "[Tuna] clearly fetched prices out of proportion to their nutritional value: there was a lively market for them."[52] Similar to a cash crop, a fish was sold for more than it was worth in nutritional terms. In other words, "fish can be sold or consumed, when times are bad, at its 'nutritional price'; but it has usually been, like the Corcyraean tunny [bluefin tuna harvested at Corfu], an asset which has maintained its producers through its exchange value when fitted into the network of Mediterranean distribution."[53] Elsewhere, Purcell suggests that eating fish was a sign of wealth and status, and supporting production a reflection of "a remarkably complex aspect of the economic hierarchies of ancient society."[54] Classical insights into the status rivalries animating sophisticated retail networks bear a striking resemblance to the treatment of bluefin tuna as "red gold" today.

Commodity empires established through bluefin tuna are not new. Her role in animating biopower in this long and crucial history is.

Recall from the Prologue that the Phoenicians developed tuna traps millennia ago along the Mediterranean and east Atlantic coasts. The traps' present-day form took shape in the Roman Christian era. It was not until Arab control of Sicily from the ninth century onward that the "traditional" or "artisanal" trap fishery with installations comprised of chambers channeling tuna into nets was established at sea,[55] its materials improved, and its operations accelerating over time.[56] During the Norman era in Sicily (about 1070 to 1200 CE), the tuna traps provided employment to commoners and wealth to feudal nobility.[57] A system of rights and of privileged access to fish ensured the Crown revenue. Regulations through decree determined: "points along the shore where bluefin fishing could be carried out, the number of boats and even the number of fishermen who would have access to the fishery," with the amounts due to the Crown, diocese, and church fixed and enforced.[58] In Christian times, fish were important to securing power: the celebration of faith happened through feast and fast, with fish a way to atone for Christ's suffering on the cross. For pagans, fish figured in sacred meals because they were associated with the depths of the underworld and thus with the ability to communicate with the dead.[59]

Records from Spain indicate that in 1552 the trap fishery produced a remarkable fifty-five thousand bluefin tuna in one year.[60] Once the source of family dynasties, making the Dukes of Medina Sidonia among the wealthiest in southern Spain, traps meant tax and royalties for the Spanish Crown.[61] Revenue from the tuna traps paid for colonial expansion overseas. Fagan claims that it was the discovery of lucrative new fishing grounds—not spices—that motivated monarchs to fund voyages abroad. To feed the devout, the poor, the military various kinds of fish on Fridays and other days of fast—and to turn common fishers into naval officers in the process—were among the ways the Dutch, the French, the Portuguese, and the Spanish practiced empire through fish in the so-called Age of Discovery.[62]

That bluefin tuna was counted at all—that is, in the preoccupation with detailing, tracking, and counting her numbers—confirms the following Weberian insight: accounting via the double-entry book was a prerequisite for capitalist development[63] and the growing ascendency of the administrative state.[64] Empire may have long organized economies but now its form takes new shape with the rise of extractive capitalism and its corresponding agent still in embryo: the nation-state.

By the middle of the seventeenth century, the same time European tradespeople shipped human cargo as slaves across the Atlantic, private merchants in the form of Genovese bankers began to invest in and control bluefin tuna traps in places such as Sicily. To pay for its wars, the Crown in Sicily sold all its tuna traps to private individuals and to the Catholic church.[65] Similar fates befell other traps across the Mediterranean and east Atlantic coasts. Yet Sicilian laws were alert to and regulated overexploitation as early as 1796, when first prohibiting the catch of small bluefin tuna.[66]

It was the Age of Liberty, when France, Haiti, and the United States declared themselves democracies, just as it was the Age of Slavery, when more than five million known Africans boarded by force slave ships bound for the Americas after July 4, 1776.[67] "As economic systems expanded in complexity and reach," Edward Baptist writes, by 1800 commoners through the deployment of merchant capital "could move goods and profits and peoples at rates and distances of time and space that had once been reserved for pharaohs." The millions of acres taken from about fifty thousand Native Americans and planted by millions of slaves structured the production of raw goods "by whip, scale, and ledger." This allowed the Anglo-Europeans to escape the impact of their own dwindling "natural resources" through violent expansion overseas. The animalization of living beings as a form of biopower was under way on the plantation, sustaining economic growth for—and giving extraordinary power to—the Anglo-Europeans in ways clearly felt today.[68]

The fish trade, then and now, is about speed and numbers: unload your catch immediately and sell it outright to brokers or user-buyers because it spoils quickly.[69] Processing and transport cost time and money, so investors began managing directly the trade of bluefin tuna bought and sold as an emerging asset class. By the late nineteenth century, capital accumulation from the tuna traps supported other interests, including wine and banking. Bluefin tuna processed and canned in olive oil emerged as a dependable way to export foodstuffs widely.[70]

By the twentieth century, as captains heard of virgin seas where fish were big and plentiful farther afield, the scale of the offshore fishing effort increased dramatically.[71] The capture of marine fish happened now that freshwater varieties were in decline in Europe.[72] After World War II, tremendous changes befell the Atlantic bluefin because much more intensive methods of capture began to dominate trade—live bait, pelagic longline, and purse seine—while production via the much older traps also expanded, if only momentarily.[73] Sarika Cullis-Suzuki and Daniel Pauly write, more generally: "In the 1950s catch from the high seas amounted to under two million tonnes; in 2006, this had grown to over ten million tonnes . . . [T]he fraction of the global marine catch originating from the high seas . . . increased from 9% in 1950 to 15% in 2003."[74] What made fishing the high seas possible was the capital investment and belief in the promise of modern technology: petroleum-powered vessels, large synthetic nets, electronic navigation systems, flake ice machines, steel hooks, monofilament lines. By the middle of the twentieth century, sea creatures bound for market could no longer fight industrialized fleets soon to be equipped with vessels so grand that they contained movie theaters, operating rooms, and onboard doctors, as if they were "floating towns."[75] The race to catch, freeze, process, and transport fish across the globe spurred a sharp increase in the size of fishing fleets in many countries seeking economic growth. Participating in this expansion were countries in Asia, Europe, and North America. Generalissimo Francisco Franco harkened back to the image of the Spanish Armada and heavily invested in his country's maritime sector to develop fisheries after Spain's devastating civil war.[76] This was to have important consequences for ICCAT.

Another nation-state desperately seeking economic rejuvenation after the devastation wrought by World War II—and U.S. nuclear warfare on Hiroshima and Nagasaki—was Japan. Of course, Japan has a rich maritime history of its own. Since the beginning of the twentieth century, Japan used ships as floating bases and focused extraction on large volumes of a few selected fish and shellfish, including salmon and crab.[77] Naval skirmishes between Japan and the Soviet Union made news headlines in the 1930s, although "few realized that behind these encounters were fisheries interests."[78] At the dawn of World War II, Japan had developed into the world's largest fishing nation by landings.[79] The combat mission forced Japan to press its great distant-water fleet into war service, which made the availability of fish protein scarce at home.

The Allied forces destroyed 80 percent of Japan's ships during World War II, with damage extending to the infrastructure of ports, communication networks, and transportation, including rail.[80] The maldistribution of foodstuffs from ongoing war and nuclear assault led Japan to surrender.[81] Families evacuated from urban areas such as Tokyo so that they would be closer to the fish supply, thus mitigating the threat of mass starvation. Theodore Bestor writes: "The postwar recovery of Japanese fishing was a priority for Occupation policy makers . . . One goal was that Japan should become self-sufficient in food production as rapidly as possible."[82] A small island nation with high population density worries about feeding its people, then and now.

Not until 1952, several years after surrender, did General Douglas MacArthur, Supreme Commander for the Allied Powers, lift the ban on Japan's fleet from fishing the high seas. Preoccupied by its inability to independently feed its own people—making access to low-priced, high-quality fish a national security interest—Japan sent factory ships first close to home.[83] Conflicts with Taiwan, the Republic of Korea, China, and the Soviet Union over rights to fishing grounds in Asia fueled Japanese ambitions farther outward into the Pacific and then across the globe. In 1956, Japanese longline trawlers first entered the Atlantic Ocean, targeting yellowfin and albacore tunas before targeting bluefin.[84] Between 1958 and 1961, official government records indicate that Japanese operations targeting Atlantic bluefin tuna grew exponentially from thirty-four metric tons to 1,385 metric tons in only four years.[85]

Georg Borgström wrote of Japan's "far-flung Fish Empire" in 1964 amid Cold War tensions: "[I]t is quite obvious that the new Japan construed its future power on a new kind of imperialism, this time of a continental, world-embracing nature . . . Her vigor in seeking fish throughout the world is impressive and ominous, as she is very far from having renounced her idea of being a great power."[86] As investment in technology made fish extraction on the high seas possible worldwide, tuna emerged as a commodity deeply tied to Japan's aspirations for empire.[87]

To state the obvious, Japan was not the only country invigorated by aspirations for commodity empires, nor was it the only one experiencing severe food shortages during World War II. The cultural importance of food insecurity as a marker of deprivation extended to other countries. The UK and Germany, for example, "went to great lengths to secure supplies."[88] When fleets returned to fishing after the war, "stocks" once in reprieve returned and sustained "excellent catches" for nearly ten years. Yet, fish "stocks" again soon declined once the pause in industrial extraction was over.[89]

Like a machine that could be engineered and fine-tuned, the new-found economy became a site through which the managerial class could self-confidently perform its powers and demonstrate its prowess in mitigating fears about food scarcity. The effort to secure the food supply finds its rationale not only in the "juridical tradition" but also in the "histor[ies] of industrial management and the political uses of technology."[90] As Foucault reminds us, concern about the risk of food shortages is forward-looking, premised on a possible event that could take place, which government tries to prevent before it transforms into reality.[91] Insecurity of other kinds has nonetheless proliferated in the effort to mitigate this anxiety. The management of risk in a futures market for tuna has, paradoxically, made the ocean itself a site of extreme volatility during the Great Acceleration.[92]

"Maximum Sustainable Yield" and the Making of Commodity Empires

The entangled histories of bluefin and fisher in the *longue durée*, from classical antiquity through World War II, allowed commercial elites in the Atlantic Ocean and Mediterranean Sea to extend their horizons of possibility for wealth and security. The bluefin became cemented into political economies, social lives, and cultures. She has become a ma-terial force through which the aspirations for empire have long played themselves out. This history has conditioned the possibility for ICCAT to become an institutional expression of, and a juridical apparatus for, the aspiration to create commodity empires from fish after World War II. ICCAT's formation coincides with, and helped to bring about, the Great Acceleration, when the rate of marine fish catches, among other trends, skyrocketed.

To understand the rapid slaughter of high-seas fish, I turn to what has become the fundamental grammar, architecture, and rationale for contemporary fisheries management: "maximum sustainable yield" (MSY). "Maximum sustainable yield" is a scientifically derived, mathematically rooted formula that, in fisheries, developed in the field of fish population dynamics. The logic of MSY is simple: to regulate the supply of fish, experts must first try to count how many creatures there are. By virtue of the social authority of number, born of the secular age, MSY provided the juridical formula for nation-states worldwide to fish as hard as possible, to the absolute threshold of collapse.[93] MSY treated economic growth from the sea's bounty as unlimited. It rationalized the accumulation of wealth and biopower sourced from the ocean. Carmel Finley writes that MSY was "locked into place by 1958," the year when the United Nations opened for signature the first conventions related to the Law of the Sea.[94] The United States flexed its trading muscle by promoting the idea of MSY. Its aspiration for its own commodity empire—then young and embryonic—expressed and enacted in this form deserves special attention.

By the 1950s, mariners from a handful of industrialized nations now chased tuna north and south, east and west, canvassing the ocean for months on end. Hawthorne Daniel and Francis Minot wrote in 1954:

> There are hooks and lines and traps; there are weirs and trawls and nets of various designs; there are harpoons, spears, and other simple weapons for direct attack, but that is all. *Or, more accurately, that has been all until hardly earlier than yesterday.* Where, only a little while ago, fishing was a comparatively simple, though a very hardy occupation, it is coming nowadays to be a highly technical business. Within the last few years, engineering, electronics and naval architecture have brought about changes that would have astonished the famous Gloucester fisherman of a generation or two ago.[95]

The unprecedented intensification and scale of the global fishing effort post–World War II corresponds with the time when Daniel and Minot declared the seas "inexhaustible," their "endless resources" still awaiting development. Alive was the sentiment of Thomas Henry Huxley, the English biologist and proponent of Charles Darwin's theory of evolution, who famously proclaimed in 1884 that "all the great sea fisheries are inexhaustible; that is to say, that nothing we do seriously affects the numbers of the fish. And any attempt to regulate fisheries seems consequently from the nature of the case to be useless."[96] Although this worldview still often proves decisive in fisheries management today, times—and beliefs—have changed.

The postwar optimism of the 1950s and the promise of modernization, of development, of "progress," fueled technological imaginations to such a degree that they had devastating, unforeseen ecological effects. Entire fisheries collapsed. The United Nations Food and Agriculture Organization (FAO) worried. Technocrats called meetings. Member states sent delegations. A complex joust between maritime states unfolded. Increased fishing power in the shadow of world war conditioned the contemporary architecture for fisheries management.

In this context, as early as 1949, the U.S. State Department formally adopted MSY as its goal in American fisheries policy, which it had already deployed when managing Japanese fisheries during the Occupation under General MacArthur.[97] Let me quote the fisheries scientist P. A. Larkin, who summarizes the contours of MSY:

> The dogma was this: any species each year produces a harvestable surplus, and if you take that much, and no more, you can go on getting it forever and ever (Amen). You only need to have as much effort as is necessary to catch this magic amount, so to use more is wasteful of effort; to use less is wasteful of food.

Proponents of MSY assumed that statistics, equations, and mathematical models would help fisheries management do its job: to find that sweet spot between catching too little and catching too much by drafting regulations that prevented catch from exceeding its limit.[98] Fish became an exploitable "resource," mere cogs in a machine, alienated from their stewards, as if sea creatures were automatons that existed solely for people. Fish were supposed to conform to, and obey, the ways in which the experts organized them based on how they bred and interacted with one another so as not to upset the "yield." The economistic motivation was clear, writes Larkin: "The real yield from fisheries is not fish, but dollars."[99] MSY was both a picture of and a tool used to sustain maximum revenue derived from a food system designed to make profit, assuring fish for people who could afford them, not for everyone who needed them.[100]

The focus on MSY in fisheries management reflected fashions elsewhere. The study of political economy was linked to physics beginning in the 1930s when the newly formed field of econometrics "imagined and modeled the economic process as a mechanical apparatus."[101] Fisheries management was not immune to this development. It too relied on physics for its explanatory rhetoric.[102] Mitchell writes that these models of explanation—with "words like equilibrium, stability, elasticity, inflation, expansion, contraction, distribution, movement, friction"—emerged not only as a picture of the market but also as instruments of investigation.[103] By the middle of the twentieth century, fish population dynamics was at the forefront of "economic biology" with its "use of statistical methods, mathematical modeling and life-table analysis."[104] The goal of fish population dynamics was to help sustain fisheries by treating exploited fish as a self-sustaining group of abstract beings persisting in geographic areas over ecological time.[105]

The United States championed the idea of MSY under the impression that mathematical rigor would make the world safe from overfishing. American diplomats urged the adoption of MSY in international negotiations, most notably in 1955 when nation-states came to Rome for the International Technical Conference on the Conservation of the Living Resources of the Sea.[106] The Rome conference provided "technical" advice on fishing in anticipation of the legally binding instruments related to the Law of the Sea Conventions from 1958. In accordance with the principle of MSY, U.S. experts in Rome assumed that fish were "resilient, stable populations that had 'surplus' individuals that could be safely harvested. There would be little harm if a few too many fish were occasionally taken."[107] Nation-states recognized MSY as the goal of international fisheries management at the Rome conference by only a one-vote margin.[108]

At the level of policy, not harvesting "surplus" meant wasted fishing effort. To maximize "yield," countries—by definition—had to exploit commercial fish to the greatest extent possible. This rationale allowed rich countries of the Anglo-European West to grow their economies by extracting fish at rates "near a crisis point."[109] It implied "that fishing could not be restricted until there was scientific proof" of a fish in decline.[110] By advocating for MSY, the United States killed two birds with one stone: it curbed the imperial ambitions of Japan and neutralized upstart countries from the developing world, including those from Latin America, because none of these countries then had either the intellectual tradition or the financial means, respectively, to develop statistical analyses.[111] At the time of the Rome conference, no institution had yet been created to provide the mechanism for realizing and coordinating MSY in tuna fish.

Nation-states have since widely adopted MSY in legally binding international instruments. These include the UN Conventions on the Law of the Sea from 1958, revised and consolidated in 1982, and the various regulatory agreements of regional fisheries management organizations such as the ICCAT convention. MSY is now the normative base, the customary standard, the bedrock for fisheries policy (both international and domestic), despite the fact that ecologists since the 1970s have repeatedly exposed this approach as inadequate to the task of "resource" management. C. S. Holling claimed in 1973: "It is at least conceivable that the effective and responsible effort to provide a maximum sustainable yield from a fish population . . . *might paradoxically increase the chance for extinctions.*"[112] In 1977, P. A. Larkin published an article in *Transactions of the American Fisheries Society* titled "An Epitaph for the Concept of Maximum Sustained Yield." He hoped his essay would discredit MSY once and for all, letting it drop to the bottom of the ocean—hook, line, and sinker—and drown.[113]

Yet MSY was always about much more than fish. The Cold War deepened the growing geoeconomic anxiety about national security and the free passage of ships over the ocean for America's commodity empire. To secure the investment in its distant-water fleet, the architect most concerned with the promotion of MSY is a little-known ichthyologist from Washington State, Wilbert McLeod Chapman. Chapman campaigned for the expansion of the U.S. fishing industry under the impression that the "equatorial Pacific [was] the equivalent of the Great Plains, with tuna as plentiful as buffalo."[114] He presumed that fish were "harvested" for utility—that is, purposed solely for people. Chapman wrote in 1955: "Fish resources cannot be stored in the sea . . . They die."[115] To not capture fish was wasteful for these "improvers." The life of a fish was for them an asset that depreciates over time, like automobiles on the newly cemented highways stretching across America.

The Great Acceleration gives the lie to the abstraction that is MSY.[116] Yet this watershed moment did not slyly arrive or float about unattended. Characters such as Chapman may not be household names, but the impact of their personal influence on fisheries policy and on the institutions mandated to achieve MSY was great. Chapman resembles other American figures who never quite make the history books. Malcolm McLean, the shipping tycoon who revolutionized global trade through containerized vessels, and Stuart Barber, the naval planner who hatched a strategy for sea power through the seizure of hundreds of islands for U.S. military bases, remind us that empires are run by real, everyday people nationalist in their aspirations.[117]

In the immediate postwar period, the United States accounted for more than one-third of the total global economy.[118] Although the United States threw its economic weight around during the 1950s, the Americans did not act alone. Reeling amid the rubble, devastation, and atrocities of World War II, countries such as England, Japan, and the Soviet Union struggled to retain or assert their own commodity empires in the war's aftermath. All wanted to maintain their legal access to fish so that they could develop and modernize their economies. These were the sites where sea powers met the Great Acceleration. ICCAT was one of the institutions that fostered this development and through which it unfolded.

ICCAT Is Born as a Juridical Echo Chamber

In 1960, five years after fisheries experts adopted MSY at the Rome conference, a group of scientists led by the Frenchman Émile Postel met to discuss the biology and exploitation of tunas in the Atlantic at a symposium in Dakar, Senegal.[119] Some scientists saw the increase in commercial catch as a threat to the artisanal trap fishery, which had been capturing bluefin in the Mediterranean basin for millennia. The possibility to fish year-round for any creature of any size caused concern among some experts, because the dominant method of capture at that point— the tuna trap—happened only on coasts in spring.[120] The rapid industrial development of the bluefin tuna fishery by longline vessel on the high seas alarmed many observers as early as the 1960s. Unease about the overexploitation of Atlantic bluefin tuna provoked ICCAT's formation from day one, so entrenched was the tuna in the project of European empire. *ICCAT was born in, and shows itself as, the crisis of overfishing.* The Dakar meeting in 1960 was the first of its kind about Atlantic fish. It happened during the year Senegal gained independence from France. An international assemblage of scientists advised the specialized UN agency, the Food and Agriculture Organization (FAO), of "the need and the urgency of creating a tuna commission for the conservation of Atlantic tunas."[121] More than a year later, in July 1962, the FAO responded by convening the World Scientific Meeting on the Biology of Tunas. More than 250 experts met for over two weeks in La Jolla, California. Once again, the summary report recommended the rapid creation of a commission with the explicit mandate of conserving tunas in the Atlantic Ocean.

The FAO then formed a working group that met twice thereafter at its headquarters in Rome in 1963 and 1965. Only fifteen people from seven countries formally participated in the first working group—known as "The Working Party on the Rational Utilization of Tuna Resources in the Atlantic Ocean." The participants came from Brazil, France, Japan, Nigeria (recently independent from the United Kingdom), Portugal, Spain, and the United States. The dominant fishing nations of France, Japan, and Spain opposed the measure calling for the creation of a commission. But a strong lobby from the United States along with Brazil, Nigeria, and Portugal supported it. Participants from the latter countries constituted a narrow majority *by one vote.* By that narrow voting margin ICCAT was born.

ICCAT's founding was thus not inevitable or without objection, even among countries with industrial fishing fleets. Rather, it was the result of negotiation between a few of the world's leading fishing nations at a time when "tuna fishing in the Atlantic did not interest many [other] countries."[122] ICCAT's rule over marine life in the Atlantic was not preordained or pursued by all states. The moment of its birth foreshadowed the dominant role the United States would play as an ICCAT member state and must be seen against the backdrop of the U.S. effort to promote relentlessly through its foreign policy the superiority of capitalism over communism at the height of the Cold War.

The executive organ of the FAO (known as the FAO Council) met later in 1963 to take up its concern with "the rapid growth of tuna fishing in the Atlantic in the absence of coordinated action to study the resources and the effect of fishing upon them." The director general concluded that action was necessary for the "conservation and rational exploitation of the tuna resources of the Atlantic."[123] More than a year later, when the working group met for a second time in July 1965, the FAO's Department of Legal Services presented draft documents in support of the creation of an Atlantic tuna commission.[124] Only nineteen people from seven countries participated in the meeting, although Spain sent a scientific observer and Nigeria an attaché from its embassy. More or less the same countries participated in this round of talks as in the last one, but Senegal sent an official delegation and Cuba, Germany, and Italy sent observers. Seven members from the Japanese delegation and six from the United States made these two countries by far the largest participating blocs.

In July 1966, with the draft document for an Atlantic tuna commission already discussed among the parties, the director general of the FAO authorized a Conference of Plenipotentiaries (COP) to convene for the purpose of establishing a commission. Invited were all FAO member states, associate members, and nonmembers (who were nonetheless members of the UN or one of its other specialized agencies) to finalize the text. Brazil hosted the conference in Rio de Janeiro. Seventeen countries formally participating at the diplomatic level adopted the treaty known as the International Convention for the Conservation of Atlantic Tunas (ICCAT convention).[125] It included a provision for the creation of a commission to meet and carry out its mandate. Not until March 21, 1969, did seven countries ratify the Rio Agreement and bring the ICCAT convention into force under international law. The drastic decline in Atlantic bluefin tuna—the fish that built empires through millennia—had awakened keen interest in a treaty now entrusted to oversee her capture.

The preamble to the ICCAT convention reads:

> The Governments whose duly authorized representatives have subscribed hereto, considering their *mutual interest* in the *populations* of tuna and tuna-like fishes found in the Atlantic Ocean, and desiring to co-operate in maintaining the *populations* of these

fishes at *levels* which will permit the *maximum sustainable catch* for food and other purposes resolve to conclude a Convention *for the conservation of the resources* of tuna and tuna-like fishes of the Atlantic Ocean.[126]

ICCAT's founding text proclaims its mandate: to bind, by contract, "governments" that share a "mutual interest" in "maximum" catch at a "level" that "sustains" or "conserves" very mobile fish "populations" as "resources." By the law of treaty—under the juridical formula of "maximum sustainable yield"—ICCAT member states must fish as hard as possible at a threshold that forestalls the wholesale extermination of a fish "stock." Similar to other marine policies and institutional arrangements, the ICCAT convention consummated the conditions for the Great Acceleration by binding and indebting parties to its predatory regime of value.

When ICCAT first convened as a commission in Rome in December 1969, delegates made real what had been negotiated so far by text in ICCAT's founding treaty. Although Brazil, Senegal, and the United States each offered proposals over the course of a decade to act as permanent host of the ICCAT secretariat (in Recife, Dakar, and Miami, respectively), the proposal from Spain won out. The high volume of tuna catches enjoyed by the Iberian country relative to other Atlantic-rim nations justified the choice.[127] In early 1970, the secretariat opened its doors in the landlocked capital of Madrid, the heart of the former Spanish Empire.

There was a compelling historical logic behind the choice of Madrid as the permanent home of the ICCAT secretariat. Spain is a nation "whose essence and destiny were intimately linked to the dominance of the seas."[128] Spain developed its economy over the last century by creating policies that would stimulate and revitalize its growth through seafaring endeavors, despite—or because of—the decline of the Spanish Armada and the formal loss of its colonial empire in 1898. Policies as early as 1909 encouraged the development of a maritime sector—shipbuilding, fisheries, ports, merchant and naval fleets—such that by the time Generalissimo Francisco Franco consolidated his power after 1939, the reconstruction of the civil and military navy became a "guarantee [of] a certain economic and symbolic grandeur" in a nation otherwise unsettled by the loss of empire.[129] A native of the town of Ferrol in Galicia Province on the northwest Iberian Coast, long home to major shipbuilding yards, Franco lobbied for Spain to house the ICCAT secretariat. At the bottom of the Seat Agreement—founded in 1970 to designate the "seat" of the secretariat in Madrid and to define the relationship between ICCAT and the Spanish government—the general's signature marks the labor involved in imagining the nation through oceanic control and military power. Of course, not all Spaniards condoned "General Franco's obsession with the glories of Spain's imperial past," evidenced by what Anthony Pagden calls the five hundred years of anxiety over Spanish colonialism.[130]

Spain does not control the ICCAT secretariat by any means, but its headquarters in Madrid is nonetheless a symbolic reminder of the place of empire in the making of the nation through the fish trade. Spain is the most powerful fishing nation in the European Union today, producing almost 70 percent of all canned tuna in the EU. Tuna is valued at 1.3 billion euros in Spain's domestic market alone.[131] (See also Appendix B.) Although Spain may feel more protected than other member states in the ICCAT apparatus because of its host role, the maritime heritage narrating Spain-as-nation nevertheless continues to erode. That narrative is succumbing to technocratic pressures emanating from a regime of value contributing to the extermination of sea creatures.

The Birth of the Global Sushi Economy

When the ICCAT treaty came into force in 1969, the market for ICCAT's signature fish—Atlantic bluefin tuna—was for canned, not fresh fish; local, not global; for low-, not high-priced tuna. Less than fifty years ago, recreational fishers in sport tournaments discarded her like common trash while bestowing trophies on the anglers who landed the biggest game fish. Atlantic bluefin tuna at the time went to canneries—not to posh sushi restaurants—or was sold for pennies a pound as fertilizer and pet food.[132] How times have changed.

For ICCAT to become an agent in the making of the Great Acceleration, additional developments were needed. We need to understand the story of how the aspiration for commodity empires unleashed the global sushi economy in the 1970s. This moment must be seen as part of lateral shifts happening elsewhere at the time—expanding the realms of profit making via finance capital; reducing production costs through cheap, flexible labor; intensifying consumer demand, and so on—out of which the new configuration of geoeconomics arose.

To compete in the Japanese fish market post-1970—given the robust investment in distant-water fleets subsidized by rich countries—tuna trap owners in the Mediterranean basin with operations close to shore reorganized production in the canneries by reducing labor costs and by streamlining operations for economic "efficiency." The firms owning the boats, the nets, the equipment increased their control over the traps—including how they were constructed and when the *mattanza* (killing) happened—in places such as Sicily where fishing activities had previously been the sole charge of the *rais* (trap supervisor). Change spurred resentment in some fishers who regarded the intrusion of monied interests as a "blatant attack" on labor and on their very, very old methods of capturing tuna.[133] I quote one: "Now there is only speculation on tuna. There is a lot of speculation . . . It is fishing for those that have money."[134] Signs of the financialization of the bluefin tuna fishery appeared. Similar to other commodified fish such as the Atlantic cod, the capture of bluefin tuna reflected organizational changes during the rise and growth of extractive capitalism.

To understand the rapid transformation in the bluefin tuna trade, we must turn to developments in Japan, where the demand for fish and the specter of food insecurity are great. The desire for the bluefin's red meat among the Japanese has a history of its own. To presume that the bluefin was a longtime Japanese cultural icon and revered traditional fare is misguided. The Japanese taste for fish varies by region: today the fatty dark red meat of bluefin tuna is preferred in eastern Japan while the pink meat of yellowfin tuna and striped marlin is preferred in western Japan.[135] Prior to the Meiji era from 1868 until 1912, the Japanese considered the best sushi the pink meat of blue marlin, not the dark red meat of bluefin tuna. Seen as a low-grade fish best suited for the working class, bluefin tuna spoils quickly, so was marinated in soy to disguise flavor with the greasiest parts of her belly reserved for cats.[136] "In fact, the sushi vendors on the streets of nineteenth-century Tokyo rarely served anything raw," writes the journalist Trevor Corson. Without refrigeration, they salted, marinated, blanched, and seared fish to keep them long enough to serve without spoiling.[137] The Japanese did not store fish in ice until 1897, when Japan opened to outside influence during the Meiji Restoration.[138]

It was during reconstruction after World War II that life in Japan radically changed, including foodways. The journalist Sasha Issenberg claims that the U.S. occupation of Japan brought greasy steaks with General MacArthur. The Japanese began to associate fatty palates with luxury. Parallel to the newly cultivated penchant for bluefin tuna arrived "foie gras, chocolate truffles, soft cheese or porterhouses-for-two."[139] Said one Japanese wholesaler, remarking on the newfound Japanese taste for fat: "It was America that raised the price of tuna, if you think about it."[140] In this narrative, supply met Japan's intensifying demand for new and distant flavors.

Orthogonal to this rendition is the political role of the Allied forces, including the United States, in promoting the fatty meat of such sea creatures as whale after World War II. Rather than conceptualize demand as the aggregation of individual taste or as an "artifact of individual whims and needs," demand in this account is better understood as a "socially regulated and generated impulse" influenced by state action.[141] As a matter of public policy, occupation authorities encouraged the resumption of whaling, in part because cetacean meat, also red, was a reliable source of protein at a time when Japan was experiencing acute food shortages. Following the war, fleets captured whales in such quantity that their meat was commonplace in "school lunches and factory cafeterias . . . [T]oday [it is] remembered by some with nostalgia for their youth, by others as an unpalatable reminder of hard times."[142] Whaling firms also helped to develop Japan's fishing industry, which profoundly shaped the nature and development of Tokyo's famed fish marketplace, Tsukiji, the center of today's global sushi economy, if only symbolically.[143]

Sasha Issenberg's *The Sushi Economy* opens with the story of what happened when JAL, Japan's national airlines, hired the diligent Canadian Wayne MacAlpine for its then undeveloped freight business in April 1971. This date coincides with the newly established ICCAT treaty. Japanese exports such as textiles, cameras, optical lenses, and other electronic devices filled airport warehouses, but planes on return flights to Japan were typically empty of cargo. Representing only

3 percent of JAL's business at the time, cargo would soon become a key component of the airline industry, transforming unprofitable passenger flights into profitable ones, and making JAL the world's largest cargo and passenger carrier by 1983.[144]

In response to a query by JAL executive Akira Okazaki about tuna fishing in the Atlantic—which other officials at JAL branches in Europe and the United States ignored—MacAlpine consulted Canadian provincial governments and informed Tokyo that sport fishers came each fall to hunt giant bluefin tuna off Prince Edward Island, which was "a source of great pride for the local government."[145] After docking bluefin, the anglers "had their picture taken with the fish and dug a hole with a small bulldozer and buried them."[146] Okazaki eventually decided to fly to Canada after some research. "He and his colleagues knew nothing about fish. 'We did not know why canned tuna was white, and raw tuna meat was red.'"[147] They discovered that the race against time to stave off decay was everything in seafood. Swift movement between the unchartered routes of supply and demand in fresh fish made for an outstanding economic opportunity—so tidily rationalized in retrospect. The newly cultivated taste for fatty meat among Japan's corporate and political elite implied that the fish they wanted most was the bluefin, the fattiest among them all.[148]

Nonetheless, the mechanism that would allow for fresh fish distribution—globally—had not yet been developed. After several trials, including ways to chill fish and transport it to cargo depots in cost-effective ways—enlisting partners, convincing bosses, and building relationships in the process—JAL built a refrigerated container, known as a "ref-con" in the airline industry. Around the Japanese Archipelago, marine life showed signs of exhaustion and the seas' pollution resulting from the high-growth decades of the 1950s and 1960s. The development of commercial refrigeration and high-speed trucking throughout Japan allowed for delivery of fresh seafood within a day. Bestor explains: "The impact on consumer tastes was profound. Old-fashioned specialties of pre-refrigeration days, such as pickled tuna in soy sauce or heavily salted, gave way to preferences for simple, unadorned, but absolutely fresh fish."[149] Entrepreneurs searched for the next frontier in bluefin tuna extraction.

On August 14, 1972, more than a year after JAL corresponded with its foreign branches, the airline flew its first "flying fish" ten thousand miles and delivered bluefin tuna from the Atlantic to Tokyo's Narita Airport en route to Tsukiji marketplace. Tsukiji dealers marveled at the feat—and the large size of the fish—and tasted fresh bluefin tuna from Canadian waters for the very first time as if they had plucked her themselves from the Atlantic. The shipment cost forty thousand dollars, a price worth the transit costs.[150] By August 1974, 91 percent of JAL's total outbound cargo from Canada consisted, by volume, of bluefin tuna.[151]

Vital to this development was Boeing's new 747 jumbo jet, which the manufacturer unveiled in 1970 with the hope of developing business in airfreight. Empty cargo planes returning to Japan at the turn of the 1970s signaled a trade imbalance. Cracks in the Bretton Woods Agreement appeared. Misaligned currencies prompted U.S. President Richard Nixon to terminate the gold exchange standard on August 15, 1971, changing the international monetary system overnight. The intent: by devaluing the U.S. dollar, American goods would become cheaper and sell abroad, generating income, which in turn would generate jobs in the United States, so the logic went. Although the World Bank and the International Monetary Fund remained intact, the collapse in the 1970s of the fixed exchange rate pegged to the U.S. dollar, backed by gold, in accordance with the Bretton Woods Agreement from 1944, "seismically shifted the balance of power, beginning a process that would gradually but inexorably empower the market at the expense of the governing state."[152] Geoeconomics went mainstream. Edward LiPuma and Benjamin Lee write of this moment: "The ultimate effect was a transfer of power from the political to the economic system, from the citizen-subject to the market."[153] By this time, the ICCAT regime was in train.

Similar to Tsing's matsutake mushroom, trade in Atlantic bluefin tuna illuminates the "shifting relations between U.S. and Japanese capital . . . [which] led to global supply chains—and to the end of expectations of progress aimed at collective advancement."[154] The "Nixon shock" and the oil crisis that followed were felt around the world, including in Japan. A report from the Japanese Ministry of Finance reads: "When U.S. President Richard Nixon took the dollar off the gold standard in 1971, it destroyed the very foundations upon which Japanese postwar economic policy had rested: the Bretton Woods system and, more importantly, the fixed yen/dollar exchange rate. This was the beginning of the end for high growth [in Japan]."[155] In response to the "Nixon shock," the Japanese revalued the yen upward on December 19, 1971.

Back at Tsukiji, a five-hundred-pound bluefin tuna at twenty-one cents per pound meant that a fisher in the United States received $105 on December 18 and $123 the day after.[156] For East Coast fishers—who had already overfished pollock, hake, and whiting by the 1960s—bluefin tuna revived a fishing industry in decline.[157] Douglas Whynott writes: "It could be said that the bluefin tuna fishermen in New England was the best example of improved trade and perhaps the most fortunate beneficiary of the economic manipulations and currency realignments that took place from 1971 to 1991."[158] To balance trade, the Japanese could be picky about importing manufactured products but they were at a disadvantage when it came to raw goods. High-priced bluefin tuna became a "weapon" in the "complex joust" over trade, with ICCAT a zone for member states to build and aspire for commodity empires big and small.[159]

The analytic point I would like to make here is not that technological leaps drive the engine of capitalist development, as if a "ref-con" or jumbo jet have agency of their own. Nor do I want to endorse a view suggesting that the calculation involved in exploiting risk is a matter of figuring out and then fine-tuning the rates of supply and demand for bluefin tuna in an already existing sushi economy. Mitchell writes: "To introduce a new technology involved defining what that technology was, which was never a merely technical question." In this case, it involved a complex convergence of people, passions, interests, and regimes of value—entangling the bluefin in the lives of recreational fishers, tackle-shop owners, local magistrates, JAL executives, Tsukiji wholesalers, truck drivers, crystalline foam ice makers, engineers, customs agents, airline manufacturers, and many others involved in the technical development, distribution, and control of goods. Exchange relied on "different attempts to introduce calculations and persuade others that they are superior to rival models and calculations."[160] Real live agents had to find, negotiate, (re)produce, service, and finance an emerging sushi economy. The lesson is not to fetishize these technologies or think that on their own they globalized the sushi economy. Acting in concert, everyday people valuing the bluefin as a commodity did.

New developments in infrastructure, labor, and communication practices implicated new geographies of power, and vice versa, under growing conditions of economic concentration. The work of producing a "ref-con" rests on deep, entangled histories—that is, the flying box depended on the collective aspirations of people already historically situated within a grid tied to developed technical infrastructures and robust political economies through which commodities flowed unevenly and at extraordinary rates, now on a global scale. The refrigerated box container enabled bluefin tuna to circulate within commodified trajectories of old. Aspirations for commodity empires reinvigorated old supply chains, with bluefin tuna the material for ICCAT member states to (re)enact cross-border politics, stoke rivalries, and forge alliances for the flow, or trickle, of monetized beings.

Although the box container provided the infrastructure to usher in the next iteration of globalization, I would be remiss to end this story without addressing the militarized context for its emergence. As Marc Levinson documents, the box container in maritime shipping was indebted to the U.S. combat mission in Vietnam, when reducing time and labor in transporting inventories overseas became intrinsic to war campaigns.[161] In an emerging geoeconomic era, the dual projects of global trade and national security were intimately linked. Without U.S. military action in Vietnam and Japan's concern over food security after World War II, the proliferation of these technologies would likely not have happened. Cowen and Smith write: "Where geopolitics can be understood as a means of acquiring territory towards a goal of accumulating wealth, geoeconomics reverses the procedure, aiming directly at the accumulation of wealth through market control."[162] Controlling the market for bluefin tuna in the image of commodity empires past and present motivates the labor and increases the burden of ICCAT member states under finance capitalism. But delegates could not have possibly anticipated this development even a few years earlier when ICCAT formed by treaty in the 1960s. A conspiracy ICCAT was not.

Conclusion: Speculative Regimes of Value

Sally Falk Moore writes: "The cumulative effect of legislative tinkering is a compound of preconditions."[163] This chapter has sought to illuminate through the lens of commodity empires some preconditions in the *longue durée* that authorized ICCAT to become an agent of the creeping extermination of the giant bluefin tuna during the Great Acceleration. Controlling almost half of the world's supply in bluefin tuna today, IC-CAT member states have a national interest in protecting their market share through risk management. To secure their wedge of the pie and to maintain their authority over the remaining spoils on the high seas, they built a new organizational form to referee economic development among competing interests. Significant here is the way a supranational regulatory agreement was able to be flexible enough to adapt to broader shifts in political economy. It had to accommodate a Keyensian model aiming to manage the market by governmental action and then incorporate a neoliberal or geoeconomic approach in which *the state as agent of finance capital not only regulates but speculates.*[164]

Across the booms and busts of globalized economies, the bluefin has been imagined, positioned, rationalized both as a cog in a profit-making machine and as an asset class to be selectively bred one day from an egg on a "tuna farm." Today she has been fancied as a derivative on a trading desk, under contract, volatile, the risk associated with her trade chopped up and dispersed throughout the ICCAT network, like a speculative security. The genetic sequence of her kin is now the target of monetization. She is like an atom split in a nuclear reactor whose fuel, in meltdown, has overheated. Read against the grain of finance capitalism, the text by Hardt and Negri offers this insight: "We believe that this shift makes perfectly clear and possible today the capitalist project to bring together economic power and political power, to realize, in other words, a properly capitalist order . . . [T]he contemporary tendencies toward Empire would represent not a fundamentally new phenomenon but simply a perfecting of imperialism." In the realm of commodity empires, at ICCAT "a new notion of right, or rather, a new inscription of authority and a new design of the production of norms and legal instruments of coercion"[165] emerged from the rubble of World War II to guarantee, by contract, the performance of an underlying entity—a fish "stock" in a futures market—akin to traders swapping currency on Wall Street.[166]

This is the world brought into being through ICCAT's predatory regime of value. It is a regime whose elements took centuries to produce. Now a fish dubbed "red gold" and an institution said to "rule the

Atlantic" have coproduced one another within a regulatory framework whose mandate to conserve oceanic life operates according to an exterminatory logic. This entanglement was un-predetermined, unpredictable, contingent. As we see in the next chapter, yet again, these decisions were made by a few definite persons, "commissioned to do exactly that," not by the "impersonal policies of drift and nonintervention," or least by Adam Smith's invisible hand.[167]

2

~~~

A "Stock" Splits

PROFITEERING *through* INTERNATIONAL LAW

Alienation inside the Beltway

In October 2010, when calls to "save" the imperiled Atlantic bluefin tuna reached fever pitch in the news media, the three commissioners from the U.S. delegation to the International Commission for the Conservation of Atlantic Tunas (ICCAT) convened their Advisory Committee.[1] Some two-dozen appointees joined more than thirty government officials in a banquet hall at the Silver Spring Hilton Hotel inside the Beltway of the nation's capital. The group included tuna wholesalers, tuna canners, tuna lobbyists, commercial fishers, recreational fishers, trade negotiators, insurance agents, marina owners, bait-and-tackle shopkeepers, environmentalists, fisheries scientists, lawyers, and policy experts from the U.S. Senate Commerce Committee, the U.S. House Committee on Natural Resources, the U.S. Department of State, the U.S. Fish and Wildlife Service under the U.S. Department of the Interior, and, not least, the National Oceanographic and Atmospheric Administration (NOAA), part of the U.S. Department of Commerce. Although this battery of experts meets every fall and spring, somehow the snaking hallways leading to the windowless room felt more claustrophobic than usual. The air was staler, the fluorescent light more artificial, the modular wall panels more intractable, the metal chairs with fake leather cushions colder to the touch.

As members of the Advisory Committee sat around a rectangular table, in recessed seating behind them were low-level bureaucrats and a handful of concerned citizens. Together in committee they reviewed new scientific developments, voiced concerns, summarized positions, and strategized how to advance their own interests via the ICCAT apparatus. At the opening of this meeting, there were hugs, congratulations, and condolences in response to personal developments that had transpired since they last met six months earlier: divorces, deaths, motorcycle accidents, newborns, marriages, the achievements of grandchildren. Some participants had been working together for a very long time. Gone were the days from decades past when security guards policed these meetings for fear they would spiral into violence as a result of infighting between fishers.

Members of the public may attend the "open session," as I did, which lasted only a half-day out of the three-day schedule. Although the public was allotted an hour to share its views, the meeting ran behind that day. A scant twenty minutes was left for a few folks to interface with a committee advising the U.S. delegation on how to regulate catch in an ocean emptying of big fish. Environmentalists from Oceana and the Pew Environment Group in addition to one commercial fisher from Plymouth, Massachusetts, issued prepared statements. No questions were asked of them. The representative from Pew called for the suspension of the Atlantic bluefin tuna fishery. Some industry representatives grimaced.

In principle, the period for public comment was supposed to act as a check on domestic administrative law, ensuring that the guardians of the sea were not making devil's bargains in the name of marine conservation. In practice, this time felt like a formality, wasted for some, a box ticked on the agenda, a concession to the demands of regulatory lawmaking, not unlike a public-service message on radio or television broadcast in the wee hours: necessary by law, short, off key, incongruous with paid programming. "We can speak 'offline'" was a commonly repeated phrase indicating that discussion was meant for closed doors.

Amid the reports and debates, the jokes and jabs, the rehearsed scripts and spontaneous ruminations, the committee's lead scientist offered his usual roundup of new research about the commercial fish under ICCAT's regulatory control. His presentation included a series of PowerPoint slides showing the tagging results of four lone Atlantic bluefin tuna caught on a pelagic longline vessel as by-catch from another fishery in the Gulf of Mexico. In May 2010, the crew tagged these four fish with devices recording information on their whereabouts— their location, their depth, the water temperature through which they swam—which was then beamed back to the scientists by satellite for, they hoped, clarity on the bluefin's migratory patterns.

By August 2010—over the course of only ninety days—four tagged bluefin tunas swam from the Gulf of Mexico down to Cuba, along the eastern Floridian coast, past the Carolinas, by New England, and up to the Gulf of Saint Lawrence in the Canadian Maritimes. One dove to depths in the Gulf of Mexico where the water column approached zero degrees Celsius. Was she on the hunt for prey, fixing to spawn, or playing with mates? The tagging results on-screen looked like Orion's Belt or some other constellation from outer space linking lines and dots to form a picture holding the human imagination in trance, as these representations have done for millennia.

Although it was "premature to draw conclusions on four fish," remarked the scientist, this moment was the only one when to me the bluefin felt alive in ICCAT's predatory regime of value, her majesty restored, no longer a servant of capital as a biological asset worth her weight in gold. Even if for a moment, attendees in the main did not zone out, fiddle on cell phones, search the Internet, whisper to a neighbor, refill a coffee cup, or grab another free Danish to pass the time. They paused and looked up at the screen, transfixed by and in awe of the vast traces this former ocean giant left in just three months while swimming in the west Atlantic. Once ninety days expired from the moment of catch and release, the tag popped up and floated to the ocean surface. What happened to the bluefin thereafter we do not know. Perhaps one of them journeyed right past a fisher in that tired room, unbeknownst to him, or jetted across the Atlantic to find kin in the Mediterranean Sea, disappearing like a phantom in a regulatory void.

A moment later, the marvel the bluefin inspired among the experts evaporated. Poof. Gone. There was an agenda to follow, a market to secure, a nation to imagine, a state to administer, an empire to amplify. To carry on the work of marine conservation relative to economic growth, ICCAT delegates transformed a fish revered for millennia into just another commodity for sale, ensuring that the soulless rhythms of extractive capitalism continued unabated. At issue was not alienation in the classic Marxian sense of "labor without interest in the product," nor was it alienation without knowledge enabling workers to "purify an inventory," as Anna Tsing describes of the matsutake mushroom.[2] To legally constitute and regularize global markets for commodity empires, ICCAT delegates must learn to disentangle and alienate themselves from the vibrant brilliance of the bluefin—of which they were fully aware—allowing them to reduce her life to a short-term investment in a futures market or, in fisheries parlance, a fish "stock."[3] Capital accumulation matters, Tsing reminds us, "because it converts ownership into power."[4] The process by which imperial powers transformed the bluefin into property to secure more and more capital is the subject of this chapter.

ICCAT Distributes Export Quotas according to Power Structures of Old

This chapter deepens the analysis of commodity empires in chapter 1 by illuminating how ICCAT legalized its predatory regime of value. It chronicles by what mechanisms ICCAT member states distributed the fruits of extractive capitalism according to power structures of old, thereby making this regional fisheries management organization (RFMO) increasingly relevant in the world of ocean governance. New is the architecture for extermination, which legally constitutes and regularizes the global market share some ICCAT member states enjoy in bluefin tuna through catch quotas. This was a development that could not have possibly been known when ICCAT formed in the 1960s.

Chapter 1 established *what* ICCAT member states regulate—export markets in commodified beings in an increasingly geoeconomic age. This chapter focuses on *how* ICCAT member states regulate their imperial ambitions by distributing risk in a futures market for prized commercial fish. I reformulate an argument put forward by Sally Falk Moore decades ago about "law as process." This allows me to situate contemporary concerns about biopower squarely within the realm of a juridical formation leveraged to control "populations" of fish on the high seas. By showing that biopower resides not only in legal ends or doctrine but in means or process, I provide an account of the ways in which a seemingly mundane regulatory apparatus contributed to producing the fish slaughter after World War II. The power residing in "law as process" must be seen as a way to profoundly shape regulatory ends, and vice versa.

This chapter sets the stage for ethnographic renderings described later in the book by characterizing ICCAT meetings as ritualized performances operating in the seam of legal outcomes, as liminal as any other ceremonial event. It is through ICCAT's rule-making process that its regulatory culture is cut and cast, confessed and affirmed, so that over time delegations return to play year after year the collective "game" of ICCAT (described in chapter 5). To see law as a ritual, a rite of passage, a "tournament of value," clarifies and supports why delegates commit to ICCAT's regulatory process in the face of disputed outcomes.[5] This is doubly true now that ICCAT and like institutions have become zones for member states to realize their "new sovereignty" in a geoeconomic age.[6] In the balance, ICCAT exercises its own authority as the primary arbiter of commodity empires achieved through the regulation of high-priced fish "stocks."

More specifically, this chapter tells the story of the moment in the early 1980s when nation-states adopted for the first time ever export quotas for bluefin tuna. Quotas began with bluefin tuna caught in the western Atlantic and were extended later to her mates in the eastern Atlantic.[7] Unlike commodities such as copper, steel, and soy, Atlantic bluefin tuna travels far and wide across maritime borders at a pace and speed technocrats cannot foresee or limit in the wild. Controlling risk in the inventory of this unbounded oceanic being came just when nation-states were finalizing an agreement on how to govern "resources" on the high seas, having extended their coastal sovereignty through exclusive economic zones (EEZs). These were codified under the revised UN Convention on the Law of the Sea (UNCLOS III) in 1982. Many nation-states now claimed proprietary rights to fish, oil, gas, and wind within two hundred nautical miles of shore—expanded from a measly twelve. Quotas for Atlantic bluefin tuna allowed nation-states to demarcate who owned what, beyond the newly forming EEZs, part of an extractive economy now scaled globally and bound legally in fisheries by the parameters of "maximum sustainable yield" (MSY) established in the 1950s. Similar to flagged vessels traversing sea lanes, the bluefin has become a "vector of law,"[8] affirming that very mobile biological assets could be enclosed for well-financed member states claiming a share of the tuna pie through the ICCAT apparatus. A handful of people from a handful of powerful member states made a handful of decisions to set new terms for the global commodities trade. It was their job.

In this chapter the reader comes to appreciate that the history of commodity empires organized the contemporary geography of supply in the world's most expensive tuna fish.[9] The length and location of a member state's coastline did not determine who claimed a share of the lucrative bluefin tuna "stock." Imperial history did. It is not surprising that ICCAT member states with the greatest share in export quota for Atlantic bluefin tuna were from, or were deeply tied to, the former metropole. Imperial powers past and present used historical catch data to justify a member state's legal claim to a wedge of the export pie in bluefin tuna. Poor(er) ICCAT member states often contested this. Quotas were more than signals to global markets about how many bluefin tuna a country may legally export each year. Quotas were also symbolic capital accrued among the member states claiming as their own valuable fish on the high seas. Quotas allowed certain countries aspiring for commodity empires to legally constitute, protect, and stabilize their investment in industries "harvesting" risky fish for export. The tragedy of extinction wrought by a predatory regime of value must be located not in the commons but in the commodity form.[10]

The Power of Law in Process

This book takes for granted that ICCAT is a legal space where member states are accountable to the norms they produce, some of them binding. An approach to law *in* society—as opposed to law *and* society—underlies my decision to foreground the values that ICCAT holds dear. I treat those values as already embedded *in* hierarchies of power at the very moment when decisions were made.[11] This approach is based on the way I saw regulatory action unfold, on the ground, ethnographically, rooted in empirical observations rather than normative assumptions.

To say that law is constitutive of, and not separate from, society allows the reader conceptually to go beyond the moral functionalist account that assumes law's central role or objective is to cope with or settle conflict. Austin Turk clarifies: "To a considerable extent, law is oriented to regulating the *symptoms* of conflict without getting at the more intractable problems of removing the *sources* of conflict."[12] ICCAT cannot on its own undo the *roots* or *sources* of conflict on the high seas emerging from, and structured by, historical relationships of imperial power and the uneven distribution of wealth established in the colonial encounter. The regulatory action of a single supranational institution operating in a dense matrix of ocean governance cannot undo or loosen the hold of exterminatory values entrenched in the dominant order of society. ICCAT member states are not equipped to single-handedly solve the overfishing problem when, in fact, they have exacerbated it.

By focusing on the deep formations of law—that is, on the power of law in process—this book adopts the sociolegal perspective that locates law within the realm of ritual, performance, culture, imagination, and representation.[13] It understands commission meetings as ritualized performances of ICCAT's regulatory culture whose dynamics render the bluefin its central token of value. The prestige the bluefin accrued as the focal point for regulatory action during ICCAT commission meetings may have appeared obvious to delegates, but this status is better understood as "betwixt and between," suspended, in need of constant affirmation when delegates (re)negotiated the terms of trade.[14] We may know who the queen will be before her coronation, but the ceremony declares and affirms it.

Time and again, I observed how delegates regulated the future of fish based on their contradictory and unsettled views of the present, whether a delegate was from a rich or poor country, industry or civil society. To treat ICCAT's rule-making process as strikingly undetermined opens up the possibility for understanding how an annual commission meeting—a ritual act—proclaimed ICCAT's regulatory culture and allowed delegates as a collective to consent to, affirm, and call into being an exterminatory regime of value prioritizing marine life as biological asset, year after year.

A cultural approach to supranational regulation as ritual act accommodates the treatment of law not as fixed and firm but as indeterminate and contingent, as an active, continuous process producing an enduring social and symbolic order, which is not to say that law is not concrete or has no force or effect or that its normative content is unimportant. Like Iver Neumann in his probing ethnography of the Norwegian Ministry of Foreign Affairs, I have observed the extent to which "diplomats have always taken pride in . . . their ability to feel their way . . . The pressure on diplomats to produce instant knowledge leads to a lot of improvisation, ad-libbing and corner cutting." "[D]iplomacy's object is," as Neumann says, "*indeterminate.*"[15] The jurist and diplomat Martti Koskenniemi also writes that "international law emerges from a political process whose participants have contradictory priorities and rarely know with clarity how such priorities should be turned into directives to deal with an uncertain future."[16] While in the field I also found "by what absurd shifts and accidents much of the law has been arrived at."[17] Law appears as a composite whole only in retrospect. Yet new rules surely relate to old ones, indebted as they are to structures inherited over the years. To say that law emerges out of indeterminacy does not imply that law is "random, accidental or arbitrary."[18]

Open for scholarly interpretation of law-as-indeterminate are at least two interlocking processes, Moore suggests. They are the "process of regularization" and the "process of situational adjustment." In the former, lawmakers try to control the extent to which they are subject to indeterminacy. They attempt to fix, harden, give form to, order, and predict social reality as best they can. The "process of regularization" produces rules and organizational forms with their own customs, symbols, and categories of representation, which may crystallize over time and become durable for the people that must deal with them. People strategize and plan their action based on what they can expect from others and on their frames of reference. To tame law's indeterminacy—to aspire for a social reality appearing stable, firm, and predictable for its makers—implies that law is not a permanently achieved state but a *process* in which social realities are constantly renewed.

In the countervailing "process of situational adjustment," Moore says, lawmakers respond to their immediate situations not by containing or limiting indeterminacies but by exploiting or generating them. In these moments, ICCAT decision makers reinterpret rules and relationships to suit their situational ends. Although I have observed both processes unfold during ICCAT meetings, for the purpose of illustration the "process of regularization" is emphasized in this chapter while the "process of situational adjustment" awaits chapter 5. There, ethnographic descriptions of ICCAT's regulatory process in action ground claims about the ritualized dimensions of law.

To characterize international legal processes as fluid, unstable, messy, and indeterminate is not to signal that legal texts are ambivalent by intention. The goal is not only to accommodate various interests or multiple interpretative frameworks, as if individual actors could know with certainty how to dictate an outcome. By moving beyond legal doctrine, I assume that law's indeterminacy—whether "of form and symbol or of content"—stems from the political process itself.[19] Law is not a neutral arbiter free of passions and interests but is fraught, always, with power and politics. Moore explains:

> Established rules, customs, and symbolic frameworks exist, but they operate in the presence of areas of indeterminacy, or ambiguity, of uncertainty and manipulability. Order never fully takes over, nor could it. The cultural, contractual, and technical imperatives always leave gaps, require adjustments and interpretations to be applicable to particular situations, and are often themselves full of ambiguities, inconsistences, and often contradictions.[20]

I regard the attempts to both regularize and exploit those "gaps" to be the very stuff of power politics transcending the ICCAT experience. The regulation of bluefin tuna catch at ICCAT commission meetings was one aspect of the labor involved in realizing the geoeconomic control over the ocean under the current conditions of international relations.

By considering international lawmaking in action—as ritualized performances emerging from a complex fluidity that expose the power of law in the temporality of the liminal—this book in later chapters offers empirical evidence about the extent to which ICCAT's decision-making process produced supranational regulations that were "adaptive."[21] To speak of regulatory production as indeterminate or liminal does not indicate a "structural deficiency."[22] Law's indeterminacy is not a problem to be solved with the proper legalese or right technical know-how. Rather, the indeterminacy inhabiting the ritualized performances of regulatory action is the exact processual requirement needed to encourage delegates—across ideologies and interests, politically left and right, rich and poor, whether representing state or industry or environmental NGO—to come back to the negotiating table again and again. Not knowing the exact outcome of the match binds delegates to the institution and encourages them to return to ICCAT's tournament of value every year.

Similar to other ritualized events, long the subject of anthropological inquiry, ICCAT commission meetings have their share of neophytes and elders, newcomers and repeat players. The "grand old men"—the delegates with ten, twenty, thirty years of ICCAT experience under their belts—have status and rank.[23] They derive their authority not only from law but from the personification of the self-evident authority of bureaucratic tradition, resonate and affirmed in ICCAT's regime of value and in the accumulated diplomatic knowledge about how to deftly play the ICCAT "game" for well-financed interests.[24] Delegates, even from the rich trading areas of the EU and the United States, expressed unease about turnovers in head delegate. Will a new captain at the helm exercise steady command at the negotiating table in the face of indeterminacy?

An important question follows: If the production of international law is indeterminate, if ICCAT rules are not predetermined, if the exact outcome is unknown at the start of each tournament, then why at the end of each meeting, year after year, does ICCAT fall on the side of the status quo in the making of commodity empires? With almost half the world represented at ICCAT commission meetings, why has there been no marked shift in who controls the global market in high-seas commercial fish? I suggest, as Koskenniemi does, that despite any number of possible choices, some were "methodologically privileged" because of the "structural biases" transcending global governance.[25] ICCAT's rule-making process was itself structured in ways that privileged some delegations over others. Delegations from rich countries more forcefully lobbied on behalf of the monied interests they represented. Their size was greater relative to poor countries, for example, allowing them the capacity to review an enormous amount of information in documents translated into only English, French, and Spanish by the ICCAT secretariat. International law in process tends toward hierarchy and domination rather than toward equality and autonomy.[26] Law can never be wholly evacuated of power politics and always holds the potential for coercion, as chapter 5 details.[27]

If the structural biases inherent in the rule-making process were more or less stacked against poor member states and served as rubber stamps for rich ones, why did nation-states invest in ICCAT at all? Why did the Pacific island nation of Vanuatu participate as an ICCAT member state when it did not have an industrial fishing fleet in the Atlantic Ocean, unlike Japan and Chinese Taipei (Taiwan)? Similarly, why did Bolivia—a landlocked country with no coastline and no ICCAT fishery whatsoever—participate as an observer in the ICCAT commission meetings I attended?[28] Why did Norway send four to six delegates to commission meetings when it has a quota for only one ICCAT fish—Atlantic bluefin tuna—but did not hunt her at the time of my field research, because she was once declared endangered and her domestic trade prohibited? In other words, without a dog in the fight, why participate in ICCAT at all?

Abram and Antonia Chayes clarify: "Sovereignty no longer consists in the freedom of states to act independently, in their perceived self-interest, but in membership in reasonably good standing in the regimes that make up the substance of international life." Nation-states realize and express their sovereignty today, in part, through their participation in and compliance with the various regulatory agreements ordering the international system. This constitutes what, since 1995, Chayes and Chayes have called the "new sovereignty."[29] The nation-state has less power to regulate and impose its authority directly over the flow of goods, migrants, capital, technology, and the like, even within its own borders. This does not mean that sovereignty has declined. Rather, sovereignty has been recast, recalibrated, reimagined as it takes new form in a supranational institution united under a managerial logic of rule. ICCAT member states put technocratic managerialism into the service of engineering "stocks" of migratory fish for economic growth, some to secure and extend their empires through the sea, as if creatures were inventories stored in the biome or manufactured for cross-border delivery.

As nation-states realize their sovereignty through compliance in organizations such as ICCAT, ICCAT in turn realizes its authority over the Atlantic based not on a power to enforce rules or police the ocean.

Instead, it is the designated forum for global elites to legally constitute and manage in concert a futures market in "stocks" of commercial fish during a geoeconomic age in which biopower is ascendant supranationally. ICCAT provides the context for each member state to consent in principle to new rules of trade, bound as they are in a dense, "tightly woven fabric" of other agreements that "deeply penetrate into their internal economics and politics." At stake is *status,* ritually derived and made through face-to-face ritual encounters. "The vindication of [that status is] the state's existence as a member of the international system."[30]

But status is something that must be performed. It cannot be assumed. Status is expressed and experienced in social relations "not through force or threat of violence," as Brenda Chalfin reminds us. Status exists through "consensual forms of rule." This understanding of status derives from an anthropological approach to the state, which "recognizes that sovereignty is deeply entangled culturally as well as socially and involves the production of meaning and ways of being and knowing."[31]

The assertion of sovereignty through a nation-state's voluntary compliance with international regulatory agreements suggests that ICCAT cannot be reduced to the simple domination of rich countries over poor ones. ICCAT is not the lapdog of the EU, Japan, or the United States. This does not mean that the EU, Japan, and the United States do not wield great influence in the ICCAT apparatus. While not the instrument of any one power or alliance, ICCAT forms part of what José Alvarez calls "the empire of law." He writes:

> In lieu of universal agreement on a single god or set of gods, we have placed our collective faith in the power of the United Nations' collective security scheme and . . . in [the economic] theory of comparative advantage. Participation and compliance with these regimes are, increasingly, the only options states have . . . Those few states outside [the UN] domain . . . may as well be *barbarians.* Not participating in these regimes is tantamount to political or financial suicide. To be brought under these regimes—to be allowed to participate in them—is to be allowed to enjoy the newly defined forms of "sovereignty" left to nation states.[32]

Participation in institutions such as ICCAT is now a requirement for rich and poor countries alike to "truck, barter and exchange" in the global commodities trade, to use Adam Smith's famous formulation. Now that the number of independent countries has tripled in the UN system since World War II, each seeks to pursue economic growth by claiming a share of the pie in shrinking "natural resources" through regimes such as ICCAT.

Broad-based compliance with ICCAT rules signals a "minimum degree of convergence of liberal economic policy and interests [which] may indeed be a prerequisite for the formation of a liberal international economy."[33] To play the ICCAT "game," the delegates representing member states must have already assumed and been enculturated in a predatory regime of value in which marine life becomes above all else a profitable biological asset for them. This is the bargain struck for peace through globalizing markets that many scholars, including Karl Polanyi, identified long ago. ICCAT member states must work together to decide how to disperse risk and share what is left of the spoils on the high seas through trade and finance. Ocean governance is a way to imagine, design, and produce new norms, authority, and jurisdiction out of legal contracts in a geoeconomic age, when global trade and national security are the central fields of regulatory application. ICCAT delegates have become what Nathanael Ali calls "techno-managerial entrepreneurs" in their own right, players in the market first and foremost, self-regulators of such market excesses as the slaughter of entire life-forms.[34]

The Rise of EEZs in the Geoeconomic Age

The rise of exclusive economic zones (EEZs) under UNCLOS III in the early 1980s profoundly affected the trade of Atlantic bluefin tuna. It necessitated ICCAT to become both a protagonist in the making of commodity empires and an enabler of the Great Acceleration. Although international law has long structured how global powers acquired commodity empires, the impact of EEZs cannot be overstated.[35] The boundaries of national jurisdiction over "natural resources" vastly expanded from twelve to two hundred nautical miles. The codification of EEZs represented "the largest claim to human jurisdiction over the globe in history." As Robert Nadelson explains, an exclusive economic zone "nominally defines jurisdiction in terms of *economics* and notionally in terms of the *resources* of the sea."[36] This moment was decisive for ICCAT, because EEZs made it possible for member states to regularize export markets by enclosing commercial fish on the high seas through catch quotas. Overnight, the bluefin transformed into the property of some member states.

Although the United States is not a signatory to UNCLOS III, EEZs first emerged in two proclamations by U.S. President Harry Truman at the end of World War II. In the first, Truman decreed the right of the United States to explore its oil and gas reserves on the continental shelf extending beyond its territorial sea. In the second, dated September 28, 1945, he asserted America's sea power when declaring the following in Proclamation No. 2668, titled, "Policy of the United States with Respect to Coastal Fisheries in Certain Areas of the High Seas":

> In view of the pressing need for conservation and protection of fishery resources, the Government of the United States regards it as proper to establish conservation zones in those areas of the high seas contiguous to the coasts of the United States wherein fishing activities have been or in the future may be developed and maintained on a substantial scale.[37]

A few years later, in 1952, the Santiago Declaration adopted by the Pacific coastal states of Chile, Ecuador, and Peru was the first international legal agreement to claim a two hundred nautical-mile limit. By the 1970s, many Latin American states extended their economic sovereignty over the sea further than nations had ever claimed before. Some countries in Africa, Asia, and the Caribbean followed suit. The rationale was clear: "The present regime of the high seas benefits only developed countries," said the Report of the Thirteenth Session of the Asian-African Consultative Committee to UNCLOS issued in 1972.[38] In response to the rapid expansion of distant-water fleets subsidized by industrialized countries, underfinanced countries adopted EEZs to protect "natural resources" along their shores. EEZs were not a conspiracy hatched in the capitals of Brussels, Tokyo, or Washington, D.C. In fact, faced with the prospect that poor states could cut off or limit access to fish and other "natural resources," rich states at the time offered proposals challenging the legality of EEZs.

After several rounds of hard negotiations from 1973 until 1982, which included the broad-based participation of rich and poor countries alike, UNCLOS III replaced the four previous Law of the Sea Conventions concluded in 1958 and put them under one umbrella. After ten years in draft, UNCLOS III opened for signature in 1982.[39] This was to be an important year in ICCAT's regulation of the bluefin tuna catch, as we will see. UNCLOS III codified what had been the customary practice of some states and extended it to them all.[40] By December 1986, of the world's 142 coastal states, at least seventy proclaimed EEZs and about twenty others had established exclusive fishery zones of two hundred nautical miles.[41] The rapid, widespread adoption of EEZs signaled their "permanent fixture" in the revised Law of the Sea in only a few years' time.[42]

Part V, Article 56 of UNCLOS III defines the parameters of an exclusive economic zone: a two hundred nautical-mile limit from which a nation-state may declare as its own marine "resources" of every kind, including fish, oil, gas, and wind. (In the 1500s, claims to the sea extended to a "cannon ball shot" of three nautical miles.) Territorial sovereignty remained unchanged at twelve nautical miles from shore. In other words, from the perspective of navigation, national defense, pollution, and other laws unrelated to extracting "natural resources," control over the high seas remained the same. New was the designation of the sovereign right of coastal states "to explore, exploit, conserve and manage the natural resources, both mineral and living," says the treaty text, and to pass domestic laws for the exercise of these rights. Likewise, "other States passing through the EEZ must have 'due regard' to the coastal State's rights and duties."[43] A foreign submarine can pass thirteen nautical miles from shore but an unannounced fishing trawler cannot.

The creation of EEZs allowed nation-states to claim for the first time in world history one-third of the ocean as their national property. Coastal zones became vast organized territories rich with "resources," some mobile, some on and in the sea, and some under the seafloor. Approximately 95 percent of commercial fisheries and nearly all exploitable mineral "resources" at the time came under the dominion of nation-states with the adoption of EEZs under UNCLOS III.[44] As of this writing, about 30 percent of the ocean is under state control.[45]

Also critical to the creation of EEZs was Article 62's articulation of the rights and responsibilities of parties for, in its words, the "utilization of living resources." UNCLOS III states: "Where the coastal State does not have the capacity to harvest the entire allowable catch, it shall . . . give other States access to the *surplus* of the allowable catch" (emphasis added). Industrialized nations did not concede their market share in fish entirely when agreeing to EEZs under UNCLOS III, as the language of surplus indicated.[46] Claims about rights to fish and about rights of access to them assured rich countries that if poor countries could not extract the "surplus," then, legally, according to the parameter of "maximum sustainable yield" (MSY), discussed in chapter 1, rich countries could.

Because slow-moving creatures such as lobster do not travel very far and remain within the boundary of a country's EEZ, economic rights to them fall under domestic law, said UNCLOS III. Other creatures swimming or flying long distances beyond and across various EEZs—such as bluefin tuna—were an anomaly. Designated under UNCLOS III as a "highly migratory species" or a "straddling stock," the bluefin along with her tuna kin and peers—shark, swordfish, marlin, sailfish—traversed the ocean beyond the newfound jurisdiction of any one nation-state. To manage "natural resources" beyond the national jurisdiction of an EEZ, UNCLOS III delegated authority over them to intergovernmental organizations such as RFMOs. ICCAT is one of them. Responsibility for the regulation of catch in high-seas fish rests jointly on the nation-states that are RFMO members, with *catch quotas emerging as a way to claim rights to fish where an EEZ could not.*

Established in the 1960s, the ICCAT convention provided its signatories the legal authority to take advantage of the increasingly lucrative sushi economy globalized in the 1970s. By the 1980s, ICCAT member states capitalized on the spatial reorganization of the ocean as a zone purposed for extraction. UNCLOS III provided the architecture for ICCAT member states to enclose the increasingly lucrative "stock" in Atlantic bluefin tuna and claim it as national property. This ability to expand—or cramp—commodity empires was now legalized not through the mechanism of EEZs but through the authority UN-

CLOS III delegated to ICCAT and its member states. Regulatory practices tightened the knot of the entanglement between the profitability of the bluefin as "red gold" and the charge of a supranational regulatory agency to conserve her "stock." This convergence was realized in the practice lying at the core of ICCAT's work: the determination of export quotas and their allocation.

The United States Angles to Split Bluefin Tuna "Stocks" in Two

Although delegates make various bets at ICCAT commission meetings each year, the most important concerns two variables: the size of the pie in bluefin tuna (total allowable catch) and which country gets how big a wedge (quota allocation). In the former, at stake is the total net number of bluefin tuna that member states agree it is possible to extract from the sea each year for export, in principle, absent IUU fishing. This involves the interests of both member states and marine advocates. Since the early 1990s, environmentalists have played ICCAT's "game" of marine conservation by launching global campaigns calling for reductions in the size of the bluefin tuna pie.

In the "game" of quota allocations, we see enacted the tension between the rich and the poor, the Global North and the Global South, the core and the periphery of commodity empires. Across the divide of wealth each country tries to maintain or increase its share of the bluefin tuna pie, however slight. The goal is to secure and grow export economies for revenue and domestic jobs as much as each country can. Environmentalists did not play the "game" of quota allocations while I was in the field.

Before unpacking the process allowing ICCAT to regularize export markets in bluefin tuna, I would like to flag what ICCAT delegates call the "two-stock theory." Since the early 1980s, ICCAT member states agreed to divide all bluefin tuna living in the Atlantic Ocean in two. Simply, fisheries managers took a map of the Atlantic Ocean, drew a straight imaginary line through the middle of it, and split bluefin tuna in half: experts designated one "stock" of bluefin tuna caught in the eastern Atlantic for the countries of Asia and the Mediterranean basin, and another "stock" of bluefin tuna caught in the western Atlantic for, then, Canada, Japan, and the United States. Although scientific evidence suggests that Atlantic bluefin tuna know no boundary on the high seas,[47] ICCAT member states adopted the "two-stock theory" even though no one theory (of one-, two-, or multiple-"stocks") was conclusive at the time.[48] The "two-stock theory" lies at the core of how ICCAT member states structure regulations related to fishing Atlantic bluefin tuna. This "project of simplification" deserves full analysis.[49]

The narrative I offer establishes the stakes in the very important bet over who controls the inventory in bluefin tuna, developed chronologically by virtue of the power of law in process, first in the western Atlantic during the early 1980s and then in the eastern Atlantic and Mediterranean Sea during the late 1990s. For bluefin tuna caught in the western Atlantic, norms produced in other international legal instruments, including UNCLOS III, influenced the ICCAT commission to establish the first quota ever for a high-seas fish, rendering Atlantic bluefin tuna a vector of law now that EEZs were fast coming on the scene of ocean governance. For bluefin caught in the eastern Atlantic, the decision to implement quotas came nearly twenty years later. Both cases illuminate the degree to which the technocrats worked through law's indeterminacy to implement norms about how to slice the pie in bluefin tuna for the regularization of export markets. Although norms such as MSY still played an important role in the next frontier of extractive capitalism, the question is not whether norms mattered or if they settled conflict, full stop, but why they were mobilized at certain moments and not others. In fact, law did not settle conflict in this case but induced and transformed it.

Throughout the 1970s, a few like-minded players agreed to regulate fishing activity for domestic implementation through ICCAT. The newly formed ICCAT commission met like a "club" over smokes and a handshake, a dynamic that has since faded, although elements of it persist today.[50] I heard that "big boys don't cry" was a jab that some ICCAT delegations had used on the floor against each other in high-stakes negotiations before my time in the 1990s. The macho tone, which makes the idea of a gentleman's club so offensive in the first place, corresponds with ICCAT's high-modernist milieu and its universalizing, abstract, biopolitical claims that reduce the life of fish to national property clustered by "population."

ICCAT member states agreed to a few voluntary measures beginning in 1975, including limits to the size of the bluefin caught so that smaller fish would stay in the sea to grow, mature, and reproduce. Meanwhile, the protracted UNCLOS negotiations were under way. Solutions were needed to stem the seizure of vessels, the confiscation of catches, the fines and fees, and the imprisonment of crews, which had been sporadic since the 1950s after some Latin American countries extended their jurisdiction over marine "resources" to two hundred nautical miles. (This was the EEZ limit that would rapidly become customary international law in the 1980s.) The United States and other countries with global fleets fiercely opposed the extended jurisdictional zones. They wanted free, unfettered access to highly migratory fish because their vessels could now reach them all. In fact, the United States encouraged its fishers to enter declared economic zones so it could reinforce its own ocean policy.[51] As I now show, the United States, without a shadow of a doubt, was the driving force behind the creation of export quotas for Atlantic bluefin tuna, just as it had been the driving force in the creation of ICCAT during the 1960s and in the adoption of MSY after World War II.[52]

At home, the U.S. Congress passed the Atlantic Tunas Convention Act of 1975, which made the ICCAT convention the domestic law of the land and delegated authority over "tuna and tuna-like fish" in the Atlantic to the commission. A year later, the Fishery Conservation and Management Act of 1976 (later the Magnuson–Stevens Act) came into law. It reorganized and consolidated control over territorial waters for the fishing industry to optimize catch. It included the provision for a two hundred–mile fishery conservation zone effective March 1, 1977. This excluded foreign vessels from catching fish in U.S. waters—but stopped short of declaring an EEZ.[53] Highly migratory commercial fish such as Atlantic bluefin tuna were excluded from the bill, but recreational ones such as billfish were not. Why a policy for one high-seas fish and a different policy for another? For several years East and Gulf Coast commercial fishers complained that the policy was inconsistent. They resented the perceived power of the monied recreational fishers who had gained protection for sport fish, such as marlin and sailfish, while for tuna they could not. The class-based, parochial politics is too obvious for elaboration here.

According to East and Gulf Coast commercial fishers, the culprit in the depletion of bluefin tuna in these years was Japan. Although Japan was required to obtain permits from the U.S. Department of Commerce to fish in American waters, they were not enough to curb catch.[54] The Americans claimed the Japanese ignored the voluntary reductions in bluefin tuna catch agreed at ICCAT in the 1970s. Why let other countries pocket cash from fish caught off U.S. waters? Said one fisher: "We are allowing foreign countries to walk away with our natural resources."[55] Atlantic bluefin tuna stoked feelings of "resource" nationalism then as she does now.

At the State Department, the United States would not budge and offer domestic protections for commercially important tunas in the 1970s for fear they would jeopardize its position during the UNCLOS III negotiations. A declassified memorandum from the U.S. secretary of defense to President Richard Nixon issued in September 1972 nicely summarizes the U.S. position after Ecuador seized an American tuna boat:

> We favor an international regime for tuna and other migratory species. Entering into an interim agreement in which Ecuador is given a preference over tuna is a marked departure from announced policy and seriously undermines our negotiating position [at the UNCLOS III] conference.[56]

Domestic legislation to protect tunas unilaterally in the United States would have also undermined the new ICCAT convention just signed into law in 1975. Like the Dutch jurist Hugo Grotius working centuries ago on behalf of his client, the East India Company, the State Department aligned with the well-organized West Coast tuna industry and promoted the modern version of the freedom of the seas. Together they advocated for international management rather than coastal state jurisdiction over fish. The canneries of Van Camp, StarKist, Bumblebee, and Chicken of the Sea; the clipper fleet from San Diego; and the lobby known as the U.S. Tuna Foundation, among others, all sought to maintain or increase access to tuna as "resources" throughout the world within the purview of international regulatory agreements—where they had a voice.[57] They claimed that highly migratory fish required management measures throughout their range.[58]

In 1980—a year before ICCAT member states adopted export quotas for western Atlantic bluefin tuna—the president of the American Tunaboat Association testified before the House Subcommittee on Fisheries about the economic importance of tuna more generally:

> [A]bout 515 million pounds of tuna and tuna-like fish from US flag vessels were landed, valued ex-vessel at about $292 million . . . canned tuna products for human and animal consumption processed in 1980 were valued wholesale at $1.2 billion . . . The Pacific Ocean has provided 92 percent of the tunas supplied by US tuna fishermen, the Atlantic Ocean only 8 percent . . . the catch of bluefin tuna of the East and Gulf States represents 1 percent of the total US catch of tuna.[59]

A billion-dollar-a-year industry during the global recessionary period of the late 1970s and early 1980s made the West Coast tuna interests in the United States powerful indeed, but this was not new either. Since 1948, commercial fishers had pressed for a bureaucratic ally in the State Department.[60] Politically unorganized at the time, the East Coast fishers left commercial representation at ICCAT to their West Coast counterparts.[61] Even though the West Coast tuna industry had little commercial interest in Atlantic bluefin tuna and conducted most of its operations in the Pacific, regulations passed domestically and at ICCAT impacted fishing activity elsewhere. Control over "natural resource" markets no matter the ocean was at stake for U.S. industry at ICCAT during this time.

Throughout the 1970s, the United States under the Ford and Carter administrations was unwilling to agree to robust, binding rules with UNCLOS III pending. Meanwhile, the "two-stock theory" had been bandied about since 1977. Recall that this hypothesis claims that bluefin tuna could be divided into two discrete aggregate groups of fish along the imaginary line of the 45th meridian west, splitting the "stock": one for the western Atlantic and the other for the eastern Atlantic and Mediterranean Sea (Figure 5).

Throughout the 1970s, the United States adopted the conventional wisdom then of one "stock." Nonetheless, the "two-stock theory" made political sense: two "stocks" meant two management programs for one

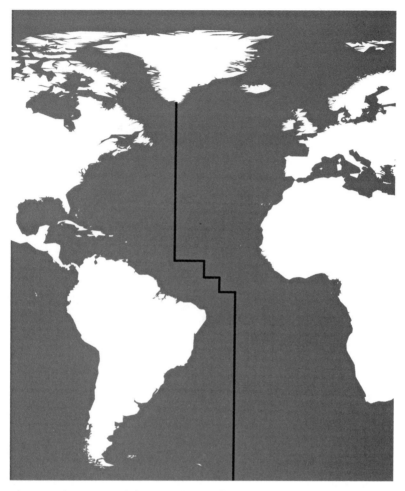

Figure 5. The "two-stock theory" splits bluefin tuna in half, along the imaginary line of the 45th meridian west, creating two "stocks" to be regulated in the west and east Atlantic.

creature, a designation that allowed for regulatory action to begin in the western Atlantic among the few countries fishing there without the many countries fishing in the eastern Atlantic having to agree to do the same. The fisheries scientist Clay Porch writes of the scientific merits of the proposal: "While controversial, most of the debate over the two-stock hypothesis does not concern its validity in concept so

A "Stock" Splits

much as the position of the boundary separating the eastern and western management units and the flux of bluefin across it."[62] Fisheries scientists believe that two distinct groups of Atlantic bluefin tuna spawn in—and return to—either the Gulf of Mexico (read: the west Atlantic "stock") or the western Mediterranean Sea (read: the east Atlantic "stock"). The bluefin returns to a place in the ocean she calls home. Yet the bluefin in her vast migrations across the open ocean meets and mixes with her kin on feeding grounds elsewhere at sea. Although the mixing of "stocks" was low enough to be considered inconsequential for fisheries management, where to draw the line in the ocean, literally, was hotly contested. U.S. fishers still complained to me years later that fishers in the eastern Atlantic caught their bluefin tuna—that is, the ones originating from the west Atlantic that migrated to the east Atlantic and Mediterranean Sea. ICCAT's scientific advisory committee began to produce "assessments" for two "stocks" in 1980, writes Porch, "where the dividing line . . . was based on the distribution of the catches and some notion of the midpoints between the continents . . . To be sure, the intent was *not* to validate the two-stock hypothesis, but to acknowledge that a reduction in fishing in the west would benefit the resource regardless of which stock hypothesis applied, provided fishing in the east did not increase."[63] A hypothesis developed in fisheries science aligned with, transformed into—and became hardened by—regulatory law. The "two-stock hypothesis" has since become axiomatic among the ICCAT delegates I met.

By November 1981, when the ICCAT commission convened for its annual meeting, the data presented by the scientific committee "was clear and irrefutable."[64] Projections showed that the fishing effort could not be maintained. Scientific advice had changed little over the years. ICCAT member states needed to act to forestall collapse in the bluefin tuna "stock" at this time.

Although inaction and lack of support for management measures by the United States had long "dismayed" Canada, what happened next sent Japan into "shock" and "trauma."[65] The United States took an abrupt about-face. Without consulting its own Advisory Committee—and without any change in the science—the lead negotiators from the United States made the following proposal: adopt a complete moratorium on the catch of Atlantic bluefin tuna, set aside a "scientific quota" for monitoring purposes, and advance the "two-stock theory." The move took members of the U.S. delegation by surprise. Without prior consultation or input, East and Gulf Coast fishers were outraged because the moratorium was "being rammed down their throats."[66] The U.S. proposal came fast, forcefully, without notice, even for members of the U.S. Advisory Committee. The memory of this blindsiding still survives in the stale conference halls described in the introduction of this chapter.

According to ICCAT Recommendation 81–1, member states were bound by rule to prohibit the capture of bluefin tuna in the western Atlantic for two years, a proposal that strangely echoes a similar one by Libya in 2010. "Scientific quotas" were limited to eight hundred metric tons for the member states actively fishing for Atlantic bluefin tuna in the western Atlantic. The measure affected Canada, Japan, and the United States but in reality Japan "had the only ox that was gored."[67] The U.S. State and Commerce Departments pressured Japan to forego a formal objection to the Recommendation—which would have otherwise exempted it from the measure—by threatening to drop permits to fish in U.S. waters.[68] Japan capitulated. On the eastern Atlantic, the proposal was palpable to countries fishing there because it included a provision prohibiting Japan from transferring its fishing effort.

The regulation of catch in high-seas tunas was now squarely in the hands of RFMOs, which the West Coast tuna industry and the U.S. State Department had angled for. Established in November 1981 were the first export quotas ever for a high-seas fish at the dawn of UNCLOS III—bluefin tuna in the western Atlantic—which legally constituted and regularized the market for at least part of the "stock" in the now emergent "red gold."

Months later, in February 1982, ICCAT met again in Miami to hammer out how to divide the "scientific quota," at which point they raised the level to 1,160 metric tons from eight hundred during closed-door sessions. U.S. House of Representatives member Edwin B. Forsythe (R-N.J.) addressed the ICCAT delegates: "Those nations which have in the past taken conservation measures to protect this resource should not now be asked to bear the lion's share of the conservation needs."[69] Although the congressman's statement was directed at the United States in the face of Japan's flagrant dismissal of voluntary measures in the 1970s, the role of the past in the allocation of quota would be more significant for the eastern Atlantic "stock," as we will see.

ICCAT member states now set in motion quotas for two "stocks," the cornerstone of Atlantic bluefin tuna management today. It is striking that the quota allocation for western Atlantic bluefin tuna among Canada (21.55 percent) and the United States (52.16 percent) has changed little since 1981, although Japan's original quota (26.29 percent) in the western Atlantic has declined over the years (Appendix B). By the 1970s and 1980s, Japan had already begun downsizing its fleet and turned to foreign suppliers for its domestic market. The declaration of EEZs forced Japanese fishers out of coastal waters while environmental campaigns—even then—provoked Japan to lower its profile and avert a negative public image as best it could.[70] Rearranging the deck chairs on the Titanic will not prevent it from sinking.

The take-home points about the establishment of export quotas for the so-called western Atlantic bluefin tuna "stock" are threefold. First, to reduce the measure to the mighty West Coast tuna industry captured state bureaucrats in the United States is too simplistic, which is not to say it is false. Although the distant-water fleets benefited from the creation of export quotas at ICCAT, the timing of the tactic cannot be explained by the power of industry alone. For several years, ICCAT delegates representing the United States had stalled the adoption of meaningful conservation measures for bluefin tuna. The quotas emerged at a particular moment, out of a broader political context, including the election of a U.S. president celebrating market values to advance national interests.

By the time President Ronald Reagan entered into office in January 1981 and became the head of federal regulatory agencies and of foreign affairs simultaneously, his administration expressed early on its "discontent" with UNCLOS III.[71] Reagan governed under the mandate of the following Republican Party platform: "Multinational negotiations have thus far insufficiently focused attention on U.S. long-term security requirements. A pertinent example of this phenomenon is the Law of the Sea Conference."[72] Under a newly elected executive, toward the close of the UNCLOS III treaty, negotiators from the United States were positioned to act differently when securing their commodity empire through the ICCAT apparatus during this time.

The Reagan administration began to treat global competition as a reason for domestic deregulatory reform. In the face of growing insecurity as a world economic power, the United States adopted market-based regulatory approaches geared to gaining a competitive edge in global trade. These approaches appeared cheaper, more flexible, more "efficient" in comparison to rules rooted in command and control. Alfred C. Aman Jr. clarifies: "Those who see the market as a means to an end in which the national government has a role to play, however minimal, are more likely to approach global issues in a manner that leaves open the possibility of new, global regulatory regimes."[73] ICCAT's enclosure of bluefin tuna through export quotas must be read against the backdrop of changes in U.S. domestic policy—the effects of which extended well beyond U.S. borders—at a time when Reagan (and his UK counterpart Margaret Thatcher) institutionalized what in some circles is called "neoliberalism." At ICCAT, at stake for the United States was not only jurisdiction over the high seas and the tuna found there. The question was how the United States would assert its sea power now that it was on the verge of pulling out of UNCLOS III. The norms emerging out of UNCLOS III shaped ICCAT's groundbreaking decision to adopt export quotas for bluefin tuna in 1981, as did the American aspiration for control over oceanic space and its anxiety about its status as a top negotiator on the world stage.

The state had transformed into an agent of finance capital in an emergent geoeconomic age. Although the United States first appeared as the champion of marine conservation at ICCAT, because it led the charge to adopt export quotas to limit catch, this position was short-lived. A year later, the United States agreed to more than double the quota to 2,660 metric tons from 1,160. By now the logic of geoeconomics reigned supreme, allowing a few member states party to the ICCAT treaty to legally claim bluefin tuna as national property relative to the global commons. To secure investment for economic growth through food production became a national security interest, itself "an emergent product not only of the political elites charged with formulating it but of the specific networks of transnational circulation in which those elites are embedded."[74] The "national interest" is not a prepackaged ideal or truism awaiting implementation but a sociolegal process made intelligible by real people subject to the convergence of historical contingencies made visible in retrospect at the bitter end.

The second lesson is not that the United States rules ICCAT, which is not to say that U.S. influence is not great. Dominant fishing nations such as Japan "chafe" under reciprocal constraints in some years more than others.[75] In some years they do not chafe at all. But long-term sovereignty expressed through a nation-state's good standing in international regulatory agreements trumps concern over short-term economic losses, even among the most powerful. This dynamic complicates the assumption that the rich always overpower the poor when in fact the rich also overpower each other from time to time. It is part of the "game."

Finally, while ICCAT meetings offered opportunities for delegates to build trust and settle conflict, they also contained the possibility of eroding trust and generating conflict. In the aftermath of the American proposal, the journalist Douglas Whynott writes of ICCAT's scientific advisory committee:

> Despite some intrigue and a few minor scandals, there had always been a feeling of camaraderie at the conference, with scientists talking and drinking late into the night. In 1982, the Japanese and American delegations kept their distance from one another, and the scientists hired by [a U.S. commercial lobby] played games of intimidation.[76]

Institutional memory lives on. Distrust within the U.S. delegation bubbles up from time to time. My conversations with East and Gulf Coast commercial fishers revealed that they still thought the bureaucrats in Washington sold off American "resources" at their expense in the early 1980s.

ICCAT member states continue to adopt binding measures for the management of western Atlantic bluefin tuna in step with the strategy of allocating export quotas for two "stocks." In 1998, the ICCAT commission agreed that bluefin tuna in the western Atlantic had become so overfished that it adopted a twenty-year rebuilding program. Yet fishers claiming they took a hit in the 1981 ICCAT Recommendation have benefited from the lucrative trade in bluefin tuna, so valuable she has become that it remains cost-effective to hire helicopters and small Cessna planes to spot overhead schools of bluefin offshore. Global positioning systems (GPS), radios (re)scrambling vessel signals, and daily market feeds by fax, now online, with auction prices from Tokyo's Tsukiji marketplace have since become staples of the bluefin tuna trade.

New Members, Old Advantage: Slicing the Pie in the Eastern Atlantic

By the mid-1990s, as growth in ICCAT membership expanded rapidly, another development in trade took root on the other side of the Atlantic in the Mediterranean Sea. First developed in Australia in the late 1980s and early 1990s, bluefin "tuna ranches" (also called "farms") heralded the next frontier in a now global sushi economy. A purse-seining vessel captures by net an entire shoal of bluefin tuna, slowly tugging inhabitants several miles before releasing them into cages in the relatively calm waters of the Mediterranean Sea. Fed mackerel, sardines, or a medley of fish twice daily, depending on weather conditions and the market price for baitfish, the bluefin circle in pens anchored to the seafloor for several months, even years, before divers, some trained military men, pluck them from cages and dealers ship them to market. The first "tuna ranch" in the Mediterranean opened in Spain in 1996. Others have sprouted in such countries as Croatia, Greece, Malta, and Turkey. In June 2011, I visited the "tuna ranch" and processing plant in Spain supplying Harrods Department Store and Nobu Restaurants. These highly intensive capital investments have exacerbated speculation in the increasingly volatile bluefin tuna "stock."[77] Financiers found at the "tuna ranches" a pot of "red gold" at the end of the rainbow. For several years, they were beyond ICCAT's regulatory control.

Although the "tuna ranches" revolutionized the extraction of bluefin tuna worldwide, they came on the scene in the Mediterranean after the environmentalists endeavored to list the Atlantic bluefin as "endangered" under CITES in the early 1990s. The European Union had become an ICCAT Contracting Party in 1997. Said one insider familiar with ICCAT since its opening days in the 1970s in an interview with me in November 2013: "ICCAT really changed when the EU came on

September 1945	U.S. President Harry Truman issues proclamations establishing "conservation zones" on the high seas off the U.S. coast, the precursor to exclusive economic zones (EEZs).
May 1955	The International Technical Conference on the Conservation of the Living Resources of the Sea (Rome Conference) adopts the juridical formula of "maximum sustainable yield" (MSY).
April 1958	United Nations Convention on the Law of the Sea (UNCLOS I) opens for signature.
March 1969	The International Convention for the Conservation of Atlantic Tunas (ICCAT convention) enters into force.
August 1971	U.S. President Richard Nixon terminates the gold standard.
August 1972	The first Atlantic bluefin tuna flies from Canada to Tokyo's Tsukiji marketplace overnight, heralding the advent of the global sushi economy.
November 1981	ICCAT issues its first catch quota for bluefin tuna caught in the western Atlantic.
December 1982	UNCLOS III opens for signature, codifying EEZs under international law.
March 1992	The first effort to list Atlantic bluefin tuna as "endangered" under the Convention on International Trade in Endangered Species (CITES) fails.
November 1992	ICCAT adopts the first bluefin tuna catch document scheme (BCD).
December 1995	The UN Fish Stocks Agreement opens for signature.
1996	The first bluefin tuna fattening "farm" opens in Spain.
November 1997	The European Union becomes a Contracting Party to the ICCAT convention.
November 1998	ICCAT adopts the twenty-year rebuilding program for western Atlantic bluefin tuna and first issues export quotas for bluefin tuna caught in the eastern Atlantic and Mediterranean Sea.
March 2010	The second effort to list Atlantic bluefin tuna as "endangered" under CITES fails.
November 2011	ICCAT adopts the electronic bluefin tuna catch document scheme (e-BCD).

Figure 6. Key developments in regulating the catch of the Atlantic bluefin tuna.

board. No longer could individual nations act on their own." It is in this context that the ICCAT commission set and allocated export quotas for bluefin tuna in the eastern Atlantic and Mediterranean Sea for the first time in 1998. Binding rules specific to the "tuna ranches" did not come until 2002.

ICCAT began regulating the eastern bluefin tuna "stock" in 1993 by limiting the fishing season and the catch levels, but it did so without quotas. Some consultants participating in the scientific advisory committee knew that their hands were tied. How could the scientists offer sound advice to the ICCAT commission when member states submitted woefully underestimated data for them to analyze? During the 1996 "stock assessment," the scientific advisory committee "expressed grave concern about the status of eastern Atlantic bluefin tuna resources in light of assessment results and the historically highest catches made in 1994 and 1995 (nearly forty thousand MT [metric tons])."[78] Even the official record—which does not include by-catch and illegal, unreported, and unregulated (IUU) fishing—indicated that the number of bluefin tuna captured from the Mediterranean in the 1990s was unprecedented. The gold rush was on.

Given the cautionary bluefin tuna "stock assessment" issued in 1998, ICCAT member states knew they had to act. With both the number of Contracting Parties and the price of bluefin tuna on the rise, member states needed to decide how they would divide the spoils and distribute risk to secure the export markets for ICCAT signatories. They agreed to put a cap on catch and divide whatever was below it. Based on ICCAT Recommendation 98-5, member states actively fishing in the eastern Atlantic and Mediterranean Sea were bound to maintain the level of total allowable catch (TAC) at 32,000 metric tons in 1999 and 29,500 in 2000. Yet the ICCAT commission exceeded its own scientific advice of limiting catch to 25,000 metric tons by several thousand, a pattern repeated throughout ICCAT's management of bluefin tuna no matter the "stock."

The politics of the pie were as contentious as those for the wedges. For bluefin tuna caught in the eastern Atlantic and the Mediterranean Sea, the European Union (read: France, Italy, and Spain) received the lion's share. Disgruntled by their allocations, Morocco and Libya filed objections based on the following breakdown (Appendix B):

China: 0.25%
Croatia: 2.97%
European Union: 63.02%
Japan: 10.00%
Korea: 2.10%
Libya: 4.06%
Morocco: 2.56%
Taiwan (Chinese Taipei): 2.23%
Tunisia: 7.27%
Other:[79] 5.54%

"Tremendous, tremendous problem . . . Allocation is probably the toughest issue at ICCAT," said one ICCAT veteran with considerable experience in ocean governance in an interview in December 2012. Characterizing these negotiations from 1998 as "hard-nosed," "tough," "tense," he recalled: "There was a lot of discussion [happening] outside of the meetings . . . People were concerned decisions were made behind closed doors . . . We were accused of too much one-on-one negotiations and I think there was a lot of that . . . [Contracting Parties] were in a bubble at that time . . . You wouldn't agree to something in swordfish unless you got something in bluefin tuna." Negotiations were never solely about just one fish, nor were they solely about just IC-CAT. Another high-ranking delegate said to me: "How can I work with you at ICCAT as I work with you [elsewhere]? . . . I may take a hit at ICCAT, but I won't take a hit elsewhere." The sentiment: Maintain peace through the regulation of trade. Save face. Ensure rank. Do not disturb status hierarchies across the spectrum of institutions in ocean governance.

According to ICCAT Recommendation 98-5, member states justified the allocation as follows: "That in order to establish an allocation of fishing possibilities, the catches of years 1993 and 1994 (whichever the higher), as laid down by [the scientific committee] before 1998, be used as reference."[80] The logic seemed straightforward enough: the extent to which a member state fished in the past determined how many fish it may catch in the future. No demonstrated history of an Atlantic bluefin tuna fishery, no quota. The larger the catch demonstrated in the historical record, the larger the quota. Yet my discussions with

ICCAT delegates did not clarify why member states agreed to focus on only those two years from the early 1990s. The documents are no help either. The more ICCAT matures, the more its documents have been sanitized of politics.

For delegates from the rich trading areas of the European Union, Japan, and the United States, the starting point in quota allocation was what they called "historical catch" or a member state's "historical performance in the fishery."[81] Whether this rhetoric reentrenched the colonial organization of the market depends on where one sat in the ICCAT apparatus. Why justify quota allocations through history and then condense all of history into one or two years, which this binding ICCAT Recommendation did? If history was the primary motivator, then why reward China with a quota at all or proportion the quota for the North African states so differently? Why reward the fleets of France, Italy, and Spain with the greatest share of quota through history when the gear historically used—the traps—was no longer the primary method for capture? Something else was going on.

Alert to attempts at regularization, poor coastal states were so discontented that they called for additional meetings to clarify the allocation criteria. From 1999 to 2001, the ICCAT "Ad Hoc Working Group on Catch Allocation Criteria" met four times to discuss and develop new benchmarks for how to distribute quota more equitably, which happened at the request of several developing countries and its leading proponent: Brazil.[82] Out of these meetings the working group produced document 01–25 titled "ICCAT Criteria for the Allocation of Fishing Possibilities." Of the twenty-four criteria listed, the first is "the historical catches of qualifying participants."[83] Now in writing, on the books, history was the primary rhetoric used to justify older geographies of supply. History—however narrow its horizon—became a legal barrier to entry in the bluefin tuna market, especially for poor member states subject to the compounding legacies of commodity empires.

Although ICCAT's primary justification for allocation was history, the politics behind it were so complex that they still shake the institution at its core. Even delegates from the rich trading areas admitted that, said one, "at the end of the day it's a horse-trading exercise." The rich had more trade and finance to leverage in their attempts to contain law's indeterminacy. The poor did not, so continually met quota allocations with resistance. Time and again poor member states angling for a greater share of the bluefin tuna quota in the eastern Atlantic and Mediterranean Sea raised the issue, and time and again the rich trading areas of the European Union and Japan defused debate. (The United States does not have a quota for bluefin tuna in the eastern Atlantic.) The extent to which discontent built coalitions and affected other areas of ICCAT's work remains open for analysis, such as when the North Africans supported fellow disgruntled Turks, and vice versa, to neutralize claims about alleged impropriety in the Compliance Committee. The irony, of course, is that when total quota size was justified in the press using the rhetoric of marine conservation, the power of law in process unleashed behind closed doors lost its name and, "unrecognized, assure[d] its victory."[84]

Conclusion: "The Tragedy of the Commodity"

Although nation-states claim "natural resources" as their own through the exclusive economic zones (EEZs) demarcating domestic waters, jurisdiction over creatures on the high seas operates according to a different logic. Nation-states could neither unilaterally assert nor recognize sole jurisdiction over high-seas commercial fish since the adoption of UNCLOS III. Nation-states that are members of RFMOs instead make rules over migratory fish in concert.[85] The regulation of catch on the high seas originates in international agreements—not in the independent assertion of state sovereignty. ICCAT's authority over marine life is global, circulatory, and extraterritorial, not national, fixed, and land-based.

Mariners had affirmed since the early modern period that ships carried their territory's law with them as they traveled through oceanic space, transforming vessels into "vectors of law" for empires to control sea lanes.[86] Similar to flagged vessels traversing shipping routes, migratory creatures such as the Atlantic bluefin tuna have become vectors of law. The bluefin has become a material force used by rich states to pioneer change in ocean governance. The difference is that while a ship carries a state's territory with it, the bluefin carries the authority of ICCAT's regulatory regime. Atlantic bluefin tuna has become not a fish without a country but a citizen of the world, a truly cosmopolitan fish, a global ambassador of ocean governance—subject to the predatory practices of ICCAT member states for commodity empires.

As Max Weber wrote long ago: "The 'legitimacy' of a system of authority has far more than a merely 'ideal' significance, if only because it has very definite relations to the legitimacy of property."[87] Although I have heard only once an ICCAT delegate refer to quotas as property, the analogy is widespread in the literature on marine policy.[88] ICCAT member states effectively claim fish on the high seas as their own when they declare or contest quotas and their allocation. As states provide the requisite property rights to contract in the market, the market provides the economic "resources" to finance state activities. ICCAT serves as the coordinating device, the "calculative agency,"[89] the bureaucratic processing zone for structuring, securing, stabilizing, and exploiting risk in the global commodities trade for commodity empires in a geoeconomic age.

It is worth asking if this case speaks to another variant of Garret Hardin's famous formulation of the "tragedy of the commons." According to Hardin's influential explanatory metaphor, the tragedy is shorthand for the inability of individuals to care for a collective "resource" because people are perceived as too self-interested, too greedy, too shortsighted when left to their own devices:

> The oceans of the world continue to suffer from the survival of the philosophy of the commons. Maritime nations still respond automatically to the shibboleth of the "freedom of the seas." Professing to believe in the "inexhaustible resources of the oceans," they bring species after species of fish and whales closer to extinction.

Hardin's solution to the problem of overexploited fish brought "closer to extinction" was simple: privatize and regulate. In his words: "[A]ugment statutory law with administrative law."[90] The allocation of property rights through catch quotas, his many followers thought, would incentivize fishers to define their own stake in the future of fish. "A pivotal moment" in world fisheries policy, the publication of the Hardin thesis hastened the adoption of EEZs under UNCLOS III because leading negotiators thought it would solve the common "resource" problem.[91] In hindsight, evidence suggests that EEZs exacerbated it.[92]

The problem is not regulation per se. It is the commoditized values ascendant in the dominant society that order and apply regulation.

The history presented in this chapter shows that an ocean emptying of big fish cannot be explained by ICCAT's "mismanagement" of the commons. The Atlantic is not a "Wild West" "where anything goes," as if fishers on some lawless frontier have pirated oceanic wealth to satisfy consumers whose pockets are deep enough to pay tremendous prices for the planet's best remaining fish.[93] While UNCLOS III may have constituted the high seas as a global commons in areas beyond national jurisdiction, ICCAT member states, led by the United States, *simultaneously* leveraged the convention as a tool of enclosure when they transformed the bluefin into property through catch quotas.[94] Bluefin tuna "stocks," both west and east, are no longer held in common. They have been privatized.

UNCLOS III delegated to RFMOs the authority to allow a member state to exert title over its global market share in fish through the catch quota system, transforming a living being into national property for the well-financed to extend their dominion over the sea. How an ICCAT member state distributes its share of quota—or *which* fishers get *how much* of a wedge allocated to them—is a matter of domestic concern. Yet, under the quota system, the bluefin has continued to decline in size and number while under ICCAT's remit. Quotas have not ensured an ocean full of fish. At issue is not a tragedy of the commons but a tragedy of the commodity form.[95] A regime of value oriented to conserving fish as property has reached its terminal point in the sixth mass extinction.

Why does the metaphor of "the tragedy of the commons" remain so intransigent? Why is the bluefin on the highs seas so available for this mischaracterization? As recently as 2008, an article from the *Economist* stated:

> If ever there were a graphic illustration of the tragedy of the commons, it is the plummeting of the world's stocks of bluefin tuna. Because they live in the high seas, these fish belong to everyone, and are thus no one's responsibility.[96]

In fact, the Atlantic variety of bluefin tuna does not belong to everyone. She belongs to only twenty-one Contracting Parties participating in the fishery (Appendix B), not all of which fished their share of quota while I was in the field. The bluefin is the responsibility of ICCAT member states. If a nation-state fishes outside of or beyond its quota, ICCAT member states reserve the right to issue trade sanctions or to lower or suspend its quota allocation. To imagine bluefin tuna on the high seas as ungoverned or ungovernable is not only imprecise, misguided, and mistaken. It dangerously reinforces the status quo by silencing the predatory values inhabiting ocean governance for the economic growth of the countries whose industries seek to amplify their commodity empires.

"It is commonplace of the classical literature on Empire," Michael Hardt and Antonio Negri remind us, "from Polybius to Montesquieu to Gibbon, that Empire is from its inception decadent and corrupt."[97] I mean to invoke this sentiment not for moral and metaphysical purposes but for juridical and political ones. Decadence and corruption are not only the absence of, or the derogation from, law. They are enabled through law. Capital rules, and it rules by law backed by state power. ICCAT member states have mastered nature's code so as to enclose it.[98] To extract the bluefin as property using MSY, until the absolute threshold of collapse, for profit, for empire, for the illusion of the never-ending cornucopia for consumption by global elites, is perfectly legal now. The magnitude of the harm wrought as a result of the lawful slaughter of big fish in only a few decades is unmeasurable. This is easily and everywhere perceived because of the self-evident enormities of the violence of extermination at a time of mass extinction. How to dislodge this mode of relating to other beings requires a close look not only at legal ends but at the ways in which power is effected and sutured to a predatory regime of value in the very process of regulatory production.

As ICCAT delegates sliced and diced fish "stocks" like derivative securities traded on Wall Street, as member states acted as agents of finance capital, distributing risk, hedging bets, contracting, speculating, profiteering, surveilling, the once giant Atlantic bluefin tuna, the one with the big heart and muscular build capable of migrating "from the west Atlantic to the east and back again in the same year," vanished from ICCAT's regulatory zone.[99] Nowhere was she to be found.

Where were the environmentalists in this story? We turn to them in the next chapter.

A "Stock" Splits

3

~~~

Saving *the* Glamour Fish

THE LIMITS *of* ENVIRONMENTAL ACTIVISM

The Savior Plot

"Bluefin tuna is a glamour fish," a boat captain from New England wryly said to me over coffee at tiny café in New York City's Chinatown in May 2011. His statement startled me, but only later did I fully appreciate its implications for ocean governance. He was right. The bluefin was meant for global elites: a delectable consumed by the leisure class, a brilliant jewel in the prestige economy, a star singled out for supranational regulatory action. She was also, as I explore in this chapter, a creature cast to play a leading role as "charismatic megafauna" in international news.[1] Celebrity was not accorded other commercial fish in the convention area of the International Commission for the Conservation of Atlantic Tunas (ICCAT). The bladed swordfish has not garnered as much public attention as the bluefin, while the plight of the other tunas under ICCAT's remit—the bigeye, the yellowfin, the albacore—rarely, if ever, made global headlines.[2]

A diva dethroned. An icon under duress. A victim of crimes against Nature. The greed of industry. The incompetence of bureaucracy. The temptation of bootleggers in black markets. Ravenous consumers freely choosing in a world of infinite goods. Environmentalists selflessly battling against the impending collapse of a species.

The bluefin, it seemed, needed a savior.

The widespread tale about environmentalists fighting to "save"

103

bluefin tuna, so prominent in popular discourse, needs to be unpacked. I want to immediately prepare the ground for a more complex account of environmental activism by confounding some common assumptions. In this introductory vignette I share an encounter I had with two environmentalists working for a major nongovernmental organization (NGO) in June 2011. It is hard to reconcile with the image of redemption.

MercaMadrid is the massive seafood market on the outskirts of the Spanish capital promoted as the second largest in the world after Tsukiji in Tokyo.[3] In the early-morning hours, well before first sunlight, I toured MercaMadrid with two environmentalists whose group seven months earlier campaigned to halt the trade of the Atlantic bluefin tuna. A moratorium on catch was their most recent demand. I traveled by car with one of them. We took the wrong road, and so, late, hurried to park to catch peak hours at one of Europe's most important food distribution centers. Did we need a pass? Neither environmentalist had been to MercaMadrid before. What transpired was indicative of the paradox of ocean advocacy emanating from the Global North today.

Like the cold innards of a sporting dome made of unadorned steel, MercaMadrid lacked the celebrity and charm of Tsukiji. After strolling through this concrete village, dodging loading trucks and puddles of water, we found ourselves in a corner windowless bar, akin to one of those dingy stops on an extended Spanish highway catering to passersby with shots of caffeine and premade processed cakes without any flavor. The floor of the café was covered in cigarette butts and empty sugar packets left behind by merchants. We pushed past the rubbish and quickly ordered coffee. The din of ceramic cups hitting their saucers subsided while business died down. Before heading out, the environmentalist who traveled to MercaMadrid on his own stopped by a stall where fishmongers competed for sales in lane after lane of stacked Styrofoam boxes kept for fish on ice.

There, one of the trained marine scientists bought wholesale two kilos (roughly four and a half pounds) of bluefin tuna—fresh from Mediterranean waters. He planned to prepare the fish later that night for his family. "Red gold" on the cheap. He smiled, and briefly hoisted in the air the meat bagged in plastic, like a champion clenching a trophy, one satisfied customer. That the vendor was considered the most egregious violator of illegal bluefin tuna catch in league with the multinational Mitsubishi seemed to escape him. "Why go on Monday? [MercaMadrid] is empty," another Madrileño, also an environmentalist, asked me days later.

How is the reader to understand the rookie purchase of a sizable chunk of fresh bluefin tuna from the foulest purveyor in the lot, not long after the NGO he represented called for a complete moratorium on catch?

This chapter considers this incident representative of a broader phenomenon at odds with dominant discourse. My intent is not at all to scold or ridicule an individual activist for his choice as a consumer. Something more complex is going on. The point I want to make instead focuses on how pervasive and normalized what I call "the savior plot" has become, so much so that even the most openly committed to reform have internalized it.[4] ICCAT is usually caricatured as a scoundrel preying on the innocent bluefin, as if the fish is a victim in a Nature under attack by insatiable appetites. In this common portrayal, environmentalists must rescue her from the existential horror of annihilation. The uncomfortable subject in need of review is that mainstream environmentalism in the Global North was blind to its own complicity in disaster. The question demanding an answer is this: what conditions explain why ICCAT's predatory regime of value has reproduced itself, undeterred over the years by a mass public communications apparatus sympathetic to the plight of the bluefin?

Framing Narratives of Overfishing

Unlike other regional fisheries management organizations (RFMOs), ICCAT has not been an unknown quantity or obscure administration in the network of ocean governance. Over the years, in fits and starts, mainstream environmentalists who had the ear of the press took their criticism of ICCAT public. They attempted to hold member states accountable for their decisions. Similar to fields such as human rights, transnational advocacy groups dedicated to the environment saw their numbers grow dramatically after the 1980s. The speed and density of their messaging, the complexity of their linkages to intergovernmental organizations, and their ability to mobilize strategic information gave environmental NGOs leverage in policy making.[5] ICCAT was fully aware of this development and was able to accommodate it.

This chapter re-creates ICCAT's recent history by identifying two discrete campaigns endeavoring, in their words, to "save" Atlantic bluefin tuna from the clutch of "commercial extinction." One focused on the western Atlantic from the 1990s and the other the eastern Atlantic from the mid- to late 2000s. Fiercely focused on one creature, both campaigns sounded the alarm about an emergency on the high seas through the rhetoric that treated bluefin tuna as a canary in a coal mine.[6] The bluefin was a creature that spoke for them all, it seemed, signaling danger from the overexploitation of a wild fish and apex predator atop the food chain. NGOs such as Greenpeace, the National Audubon Society, Oceana, the Pew Environment Group, and the WWF (World Wide Fund for Nature) figured prominently in developing transborder alliances in ocean governance.[7] ICCAT member states were acutely aware of the commission's vulnerabilities and worked hard to meet and redirect the challenge to their authority posed by environmentalists.

Environmental campaigns launched to stem the tide of overfishing the Atlantic bluefin tuna must be seen as part of the regulatory process of "jawboning." This, Abram and Antonia Chayes write, is "the effort to *persuade* the miscreant to change its ways."[8] Circulating narratives that name, blame, and shame lawmakers may induce at least some degree of regulatory compliance. Unlike domestic law, international agreements are without the stick of a navy, the authority to collect tax, and the formal mechanisms of representative democracy. ICCAT does not have a fleet of warships or a taxman, nor does it hold regular elections by the direct vote of citizens. And yet, as Chayes and Chayes observe, "It is remarkable that lawyers and international relations scholars, whose everyday stock-in-trade is persuasion—including persuasion of the decision makers—should pay so little attention and, by implication, attach so little significance to the role of argument, exposition, and persuasion in influencing state behavior."[9] This chapter explores the preexisting, persuasive frames available to the publicists who characterized ICCAT in one way—but not in others. ICCAT, it turned out, did not have to change itself drastically. It already contained within itself the mechanisms for deflection and indifference to commodification as a decision-making body whose predatory regime of value remained dominant in the rest of the policy-making world.[10]

The *New York Times*—the newspaper of record in the United States—provides a window through which to discern which frames circulated in popular discourse; why and when there was active uptake of these frames in the press; and what effect these frames had on ICCAT and on fisheries management more generally. Given the continued decline in bluefin tuna over the years, the media frame on offer emphasizing regulatory failure appeared self-evident. ICCAT was negligent and incompetent; its member states turned a blind eye to illegalities on the high seas; industrial fishers cashed in on too much "stock." This rhetoric was so compelling that environmentalists could periodically mobilize and circulate it for regulatory effect.[11] But what other frames had to be sacrificed to make room for this one?

Work by Erving Goffman is instructive. It recognizes that frames about overfishing, for instance, exist prior to actual news events. Frames shape an audience's subjective involvement in the news. For Goffman, frames are conventionalized. By definition, frames bracket how to interpret events, making them a mode of containment in their own right. People bring to a situation a particular understanding of the world that channels how they make sense of their experience, based on memory, social assumptions, and background knowledge. Frames inform and precede the stories journalists write about, "determining which ones reporters will select and how the ones that are selected will be told." Journalists do not document just the facts. They also recycle typifications. Goffman clarifies: "What appears, then, to be a threat to our way of making sense of the world turns out to be an ingeniously selected defense of it. We press these stories to the wind; they keep the world from unsettling us." The "hundred liberties" taken by the tellers of "frame fantasies" replay beliefs about the way they think the world works.[12] The status quo can be difficult to challenge through conventional news outlets because the frames adopted to circulate stories must already be prepped as legible, intelligible, enjoyably persuasive, not dissonant for a readership on the go.

Always receptive to the theatrical, Goffman entertains the idea that news is not only about providing information. It is also about presenting dramas to audiences. Environmental campaigns covered by the press rely on their high capacity for drama. This dynamic exacts the corporatized media environment in the United States, of which the *New York Times* is a part. Privatized news outlets need revenue from advertising dollars to stay in business.[13] Communications industries, here in combination with environmental NGOs, sell drama to audiences. This does not necessarily imply that environmental catastrophe is overstated. The frame of the endangered species—especially if charismatic megafauna—makes sense to an audience primed to hear a familiar plot in a social drama with a martyr on a mission to rescue an innocent victim from a nasty villain. By contrast, the frame of the acidic ocean, as just one example, has no ready-made script. Its threat to planetary futures is enormous, but readers have no context or bearings yet allowing them to narrate this story in a way that includes their own imaginative, civic, or activist participation in oceanic futures.

The dominant frame on offer in the news media went like this: ICCAT was flawed and inadequate to the task of marine conservation, broadly conceived. This frame reinforced, paradoxically, the worldview imagining that a fish systematically destroyed could actually be "saved" by ICCAT member states. The depleted "stock" appeared, narrowly, as a technical or management "problem" to be "solved" through the regulatory action of one lone institution. This truth borrows from old terrain. Faith in the secular notions of "progress" made possible—and doubled down on—the expectation that a global trade organization mandated by treaty to secure and regularize the export markets of member states could protect and restore plenty of bluefin tuna now worth their weight in gold. With enough prodding, with enough regulatory tinkering, with enough surveillance by environmentalists, the frame implies, ICCAT member states could "save" the bluefin from overexploitation and replenish the inventory for elite consumption. Those most outwardly committed to reform—environmentalists—played the role of heroes in the public imagination because that role fit the available scripts of "progress" recited in high modernity for anglophone audiences in the Global North. This allowed everyone to conceal from themselves their own complicity in brutal domination.[14]

Important are also the frames eclipsed from view by the dominant narrative. In the words of Arturo Escobar, "a certain order of discourse produces permissible modes of being and thinking while disqualifying and even making others impossible."[15] The dominant frame left untreated, unproblematized, and unimagined more systemic critiques. Quota allocation schemes, for example, rarely saw the light of day in the press, even though delegates from poor coastal states voiced this concern in ICCAT meetings. Lively popular discourse emphasized instead the "game" of marine conservation (how big should the bluefin tuna quota be?), not the "game" of economic growth (which country gets to export how many tons?). The absence of quota allocation in the news media indicates the extent to which mainstream environmentalists spoke past ICCAT decision makers. In the media blitz, or what Edvard Hviding brilliantly calls "biodiversity rescue operations," the savior plot paradoxically reinforced the very problem the environmentalists claimed to be combatting.[16] The active uptake of the savior plot extended and intensified, rather than minimized and contracted, a predatory regime of value. It prevented debate about a regulatory apparatus that, since its founding, functioned on behalf of wealthy elites in the Global North who as the beneficiaries of empire were allowed to publicly mourn the loss of what they systematically commodified, consumed, and destroyed. Renato Rosaldo calls this phenomenon "imperialist nostalgia."[17]

I begin the chapter with a few words about my approach to frame analysis, and then dive into the first organized campaign to "save" bluefin tuna in the early 1990s. I offer context for understanding why the emergence of the savior plot—however ahistorical, convenient, predictable, and narrow in scope—was qualitatively different in the second campaign in the mid- to late 2000s. While both campaigns yielded concrete policy reforms, the limits to popular resistance were palpable. The dominant frame obscured how constrained in their critiques environmentalists really were when ICCAT member states met by commission. Headlines loosen their hold as the chapter progresses, allowing for stories not told in the news media to emerge through ethnographic ways of knowing. The critique of ICCAT rooted in the commingling of "geoeconomics,"[18] sea power, biopolitics, colonial histories, commodity empires, and finance capitalism remained outside the reach of conventional frames. This continues to be the case even as the ocean grows ever more stressed at a time of mass extinction. I conclude with perspectives rooted in environmental justice, and suggest that main-stream environmentalism, in its current form, is not the revolutionary corrective required to counter the contemporary order of things.

"Stocks" Go Public

I want to get something immediately out of the way. The question about whether or not environmental campaigns had an impact is not a very probing one. Of course, environmental campaigns had an effect. Anxiety about the looming disappearance of bluefin tuna forced ICCAT out of the shadows and, in response, compelled member states to issue conservation measures constraining catch, some binding. Without question, environmental campaigns yielded results for their constituencies—at least in the short term. Figure 7 illustrates that the spike in regulations issued for Atlantic bluefin tuna corresponded with sustained coverage of ICCAT in the news while environmentalists campaigned beginning in the early 1990s. This is important. Preoccupation with the bluefin is registered both discursively in the news media and administratively in the regulatory decisions of ICCAT member states. This is not a coincidence.

Public demand for ICCAT to pull back on "harvest" came in waves and left its mark. Beginning in the year the ICCAT treaty entered into force in 1969 through the last ICCAT commission meeting I attended in 2012, I found that in the forty-seven stories mentioning ICCAT in the *New York Times,* 89 percent discussed or took as their headline the "dwindling" or "imperiled" Atlantic bluefin tuna. Nine in every ten stories associated ICCAT with the bluefin in this prominent news source. After 2005—during the second campaign to "save" Atlantic bluefin tuna—this figure rose to an astounding 97 percent. Virtually all of the thirty-two stories in the *New York Times* from 2005 through 2012 linked ICCAT with this one fish.

No wonder a fisheries manager from the Mediterranean basin expressed this frustration: even bureaucrats from his home ministry assumed that ICCAT was dedicated to regulating the catch of only the Atlantic bluefin. Authoritative narratives communicated—and helped to create—a codependent relationship between the bluefin and ICCAT. Atlantic bluefin tuna became a material force shaping ICCAT's agenda,

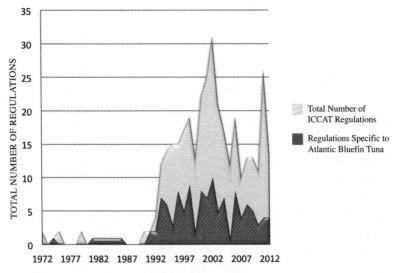

Figure 7. The number of management measures the ICCAT commission issued, both binding and nonbinding, spiked in the early 1990s during the first campaign to "save" bluefin tuna.

directing its everyday work, creating organizational demands, transforming the bluefin into a model on which the future management of other commercial fish now depends. In turn, ICCAT has contributed to the making of "red gold" and the precarious fate of a sea creature no longer permitted to grow as big as she once did.

Although environmentalists perform various roles in supranational rule making, this chapter focuses on the communicative dimension of their work because their narratives were not only world-making.[19] They were ICCAT-making. As environmentalists developed and re-proposed their own language of self-validation, they contributed to and shaped the horizon of bureaucratic possibilities in the ICCAT apparatus.[20] Before detailing these developments, I want to clarify my approach to frame analysis.

Why the *New York Times*?

To be clear, this chapter does not detail how journalists collected and crafted their stories, or how NGOs strategized and formulated their campaigns. Instead, it reads within and beyond the cultural frames available to anglophone audiences. These frames are not necessarily in the control of their makers.

My search of ICCAT in the *New York Times* stretched back to 1969—the year the ICCAT convention entered into force—which allowed me to chart coverage over time and remain alert to change in frequency and kind of story.[21] A study commissioned by Human Rights Watch of policy makers in the European Union and the United States confirmed that, indeed, most of them read the *New York Times* (and checked the BBC throughout the day). Matthew Powers, in a study of NGOs in the realm of human rights and humanitarian action, learned from his interviewees that "political elites tend to see media coverage as a proxy for public opinion." Said one interviewee: "[Government officials] can't ignore you when you're in the *New York Times*." Said another: "The *New York Times* is very influential at the United Nations. So if you really want the United Nations to sit up and pay attention to something, it's actually very helpful to have it in the *New York Times*."[22] Other news outlets commonly source and reproduce material from the *New York Times* in print, online, and over broadcast television on local, regional, national, and international scales, rippling the reach of this news brand.[23]

Variability in mandate, strategy, leadership style, membership, history, and, not least, funding suggests that environmental NGOs were not a "monolithic bloc."[24] Even so, NGOs cared a great deal about the prestige afforded them when appearing in prominent news outlets such as the *New York Times*. Coverage may catch the eye of their donor base and confer them credibility as players in international environmental politics. Both are critical to the survival and growth of NGOs. Alternatively, representatives from state and industry learn from media coverage what the demands of environmentalists are. Over time, NGOs have come to occupy a space "socially proximate to journalism," with the former seeing the latter as an ally in the pursuit of publicity.[25]

At ICCAT, NGOs served two primary roles. As *experts* in law, trade, and/or marine science, many with PhDs, environmentalists appeared trustworthy, informed, experienced, their claims based in evidence. As *advocates*, environmentalists raised public awareness about important issues in marine conservation through campaigns and arresting visuals.[26] Although the relationship between environmental NGOs and IC-

CAT member states was at times adversarial, I found their critiques of ICCAT more probing in person than in the press. As watchdogs, they appeared to me more pug than pit bull in the public sphere. This is because the narratives that went public conformed to the ordinary, humdrum beat of the *New York Times*, a corporate media brand susceptible to falls in revenue and averse to extreme departures from the status quo. This media complex was vulnerable to reproducing the simplistic savior plot, which sells drama but, in doing so, blinds readers from complicated histories and messy paradox. I suspect this complex limited what the journalists could publish too, even if unknowingly.

NGOs were important sources of information for reporters who needed to remain in the know for journalistic credibility. Unlike the International Whaling Commission (IWC), ICCAT at the time of my research prohibited journalists from attending its meetings, except during the opening and closing plenary sessions when reporters dashed in, flashed cameras, and brandished recorders for a sound bite.[27] These moments—mere photo opportunities—bookended the annual performance of the ICCAT, at best a gesture of openness offered to the scribes dominating the institution's public image. Yet, as one ICCAT veteran told me in December 2012, without talks behind closed doors, delegates feared they would be "blasted in newspapers," which would erode trust in negotiations and make "the whole thing fall apart." Indisputably, ICCAT delegates from across the political spectrum were acutely aware of the institution's appearing negligent to outside observers. ICCAT's "information wars," as one delegate characterized them, intensified over time and became operatically dramatic, as we now see.

On the Campaign Trail: First in the Western Atlantic

The first time ICCAT appeared in the *New York Times* was May 1, 1973. A story in the sports section reported on bluefin tuna tagging results for the newspaper's audience of recreational fishers along the Atlantic seaboard. The prominent scientist Frank Mather of the Woods Hole Oceanographic Institute in Massachusetts expressed concern about the "dearth" of bluefin tuna "because of the long-standing exploitation" of young fish. Mentioned in passing, the ICCAT convention had entered into force only a few years earlier. ICCAT's discursive formation included from the get-go the overfishing of bluefin tuna and the merit of scientific expertise.

Almost a decade later, with the mining of "red gold" now under way, ICCAT appeared again on March 7, 1982, albeit regionally in the New Jersey section of the *New York Times*. This occurred months after the commission—in a surprise move—first issued catch quotas for only bluefin tuna caught in the western Atlantic. (See chapter 2.) "The Tuna Industry Is Looking at New Jersey as a Crucial Test Case" read the headline. Even decades ago, fears about "alarmingly low," "highly endangered" bluefin tuna, said the article, signaled impending catastrophe. Without coincidence, this story appeared in the section dedicated to the home state of Republican congressman Edwin B. Forsythe. Forsythe played a key role in the negotiations when ICCAT first adopted export quotas for bluefin tuna.

Not until August 22, 1986, did ICCAT hit the main section of the *New York Times*. That day an in-depth cover story appeared titled "Taste for Sushi Spurs Rush to Hook Tuna Off L.I. [Long Island]." Photographs of bluefin tuna on docks in Montauk came at a time when capital flows from the Japanese postwar economic "miracle" seemed to threaten American power. The story used a disturbing, chronic, orientalist frame. The problem of overfished seas emanated from "other" voracious consumers in Asia, it seemed, not from commercial fishers from industrialized countries who profited most from supplying the market in "red gold." These countries included Canada, France, Italy, Japan, Spain, and the United States, among others. The narrative of excessive consumption offered to anglophone audiences a familiar frame rooted in demand and cultural appetites, not in supply and regulated extraction. The tension between the celebration of a field of choices in a world of endless goods and the indictment of consumers for the impact of their choices fit the frames of—and made sense to—bourgeois readers ill-informed about the way they get their food.

In the early 1990s, news coverage of ICCAT spiked. Elevated reporting corresponded with the first organized campaign to keep Atlantic bluefin tuna from becoming the next "buffalo" on the "rolling blue prairie." The campaign targeted the duplicity of delegates "blowing smoke" at ICCAT meetings.[28] A growing chorus of environmentalists— led by the prominent ecologist Carl Safina—sounded the alarm in the western Atlantic. Bluefin tuna captured by Canada, Japan, and the United States continued to decline despite the effort to privatize the "resource" through the adoption of export quotas a decade earlier. An avid fisher and PhD directing the National Audubon Society's Living Oceans Program, Safina is an influential, recurring figure in environmental campaigning about bluefin tuna. Safina with his scientific credentials appeared a trustworthy whistleblower capable of taking on ICCAT's vested interests.[29] The writer Douglas Whynott explains: "The environmental movement had become involved with bluefin . . . spearheaded by the Audubon Society" under Safina's influence as a nature lover witnessing firsthand the decline of marine life.[30] He wanted to show that fish are wildlife too—a radical proposition at the time. On the heels of the global campaigns to "save" the mammalian whale and dolphin, Safina considered the candidates for what he called a "flipper with gills" to serve as an emblem for why many fish have gone.[31] He found in bluefin tuna a fish with a rich scientific literature, a deep history, a high market value, one managed by a "negligent" bureaucracy traveling by airfreight to meet on the other side of the globe a Japanese auction house ready to pay thousands of yen for just one fish.[32] Together these elements made for a "contentious campaign." It was a compelling story. Safina and the National Audubon Society worked to get it in the news.[33] They helped to put the bluefin on the map for anglophone audiences in the west Atlantic who in the main were unfamiliar with her at the time, unlike the Europeans and North Africans, whose history with her ran deep, as chapter 1 details.

There is an immediate point to make about Safina's dual role as an expert-advocate. I do not want to lose sight of the fact that when credentialed scientists speak their truths they do so at considerable risk to their careers and reputations, which lends an aura of authenticity to their claims.[34] Discourse about the climate crisis, also mired in science, expresses a similar dynamic. Even so, I want to resist the common characterization treating Safina as a cowboy environmentalist ready to fire at his enemy or as a lone hero who took on the unsavory operators of state and industry at once. While I do not want to erase or downplay his agency either, I want to consider the narrative conditions that enabled him to appear as a whistleblower in the first place. The story about stressed bluefin tuna in the western Atlantic came at a time when the Canadian government issued a moratorium on fishing the once ubiquitous northern cod, which prompted Spain to send its navy across the Atlantic to protect fishing fleets in another "cod war." (Commodity empires clearly animated this development too.) At this time dolphin had just achieved protection in the United States through the U.S. Dolphin Protection Consumer Information Act (DPCIA) of 1990, itself emerging from the campaign for the labeling of canned tuna as dolphin-safe. Stories about threats to marine life had already been told elsewhere, and thus had traction in the press.

In retrospect, the campaign to "save" bluefin tuna clustering in the western Atlantic bears all the hallmarks of the later one targeting ICCAT's management of bluefin tuna in the eastern Atlantic and Mediterranean Sea beginning in the mid-2000s. Those hallmarks included: the insatiable appetite of Japanese consumers, the delegates with deep industry ties, the commission that ignored its own scientific advice, the quotas that exceeded the directive of "maximum sustainable yield," and, not least, the threats to list Atlantic bluefin tuna under the Convention on International Trade in Endangered Species of Wild Fauna and Flora (CITES). Regarding the last, a *New York Times* headline from September 17, 1991, stated: "Appetite for Sushi Threatens Giant Tuna." In a cover story in its weekly science section, the article reported: "The National Audubon Society has taken the highly unusual step of formally proposing that the bluefin be listed as endangered, and that the international trade in the species be banned." The article noted that it was "the first time that a commercial fish has been proposed" for a CITES listing.

Because only nation-states party to the CITES treaty may propose a trade ban, Safina first lobbied his native United States for one. The United States declined the pitch. Safina later found an ally in a Swedish colleague from the WWF, Lennert Nyman, who was also a member of the National Swedish Board of Fisheries. Not a signatory to the ICCAT convention, Sweden—"whose own once-productive bluefin fishery had vanished"[35]—ultimately submitted a proposal to ban Atlantic bluefin tuna from international trade in preparation for the 1992 CITES meeting. Threats of a CITES listing spurred ICCAT member states to act out of fear that CITES would capture and kill ICCAT's authority over bluefin tuna and, possibly, other high-seas fish. It was a brilliant strategy repeated eighteen years later when Prince Albert II of Monaco proposed an Atlantic bluefin tuna trade ban through CITES in 2010, even though both efforts failed. (Monaco was not a signatory to ICCAT in 2010.)[36]

Regarding the proposal from 1992, the journalist Charles Clover writes:

> The 1992 proposal was withdrawn [by Sweden] after intense diplomatic pressure from Japan and internal political pressure from tuna lobbyists within the United States. There is no record of what the Japanese did to the Swedes to bring about this withdrawal, whether it was threatening not to sell them Toyotas or Sony TVs, or perhaps erecting more tariff barriers to Volvos. But it worked.[37]

According to Safina—who attended ICCAT meetings—also at play was Spain's attempt to keep Sweden out of the European Community unless it dropped the CITES proposal.[38]

While the failure to ban the trade in Atlantic bluefin tuna through CITES in 1992 and 2010 is laced with diplomatic intrigue and intense arm-twisting, the proposals from both years bore similar fruit at ICCAT. In the first CITES campaign from the early 1990s, ICCAT member states agreed to reduce the quota for bluefin tuna in the western Atlantic by 25 percent over a four-year period, reflected in ICCAT Recommendation 91–1. Although the campaign to halt the bluefin tuna trade entirely was dead in the water, the ICCAT measure was a "hard-won victory for conservation advocates."[39] Environmentalists appeared to win one battle in a protracted information war over the besieged bluefin tuna.

A year later the commission adopted ICCAT Recommendation 92–1 in November 1992, which included the proposal for the groundbreaking bluefin tuna "statistical document program," curiously reissued in the second campaign some twenty years later. The initiative created an international paper trail by recording—for each bluefin tuna—the exporter, the importer, the country of origin, the area of harvest, the gear used, the weight, all for customs control—like a passport for a chunk of fish. The original proposal to adopt a "certificate of origin" by Canada, Japan, and the United States was replaced with the language of a "bluefin statistical document" following "intense negotiating" by members of the European Community. The change in language signified a change in legal obligation—making the revised version less stringent—because it "allow[ed] [the] use of a Commission-accepted document or its equivalent, instead of a certificate that would have required validation by a government official." In other words, ICCAT member states delegated to themselves the international authority to affirm or deny Atlantic bluefin tuna entry at port based on how they interpreted the former General Agreement on Tariffs and Trade (GATT), the precursor to the World Trade Organization (WTO).[40] The aim of the "bluefin catch document" (BCD) was twofold: mitigate illegal, unreported, and unregulated (IUU) fishing and better count the number of fish at sea for "stock assessments" by scientific committee.

To monitor catch required technical skills and investment, so ICCAT created the commission-wide "Permanent Working Group" (PWG) to oversee the effort by committee (Appendix C). Discourse about the (mis)management of Atlantic bluefin tuna spurred a notable change in ICCAT's organizational structure, extended the commission's authority over compliance, and birthed new methods of regulating the global fish trade. Atlantic bluefin tuna had again become a "vector of law" on the high seas and pioneered what is considered an innovation in fisheries management through the adoption of BCDs.[41]

ICCAT member states indeed responded to public calls for regulatory action in the early 1990s. Shortly thereafter Safina ridiculed ICCAT in an issue of *Conservation Biology*, playing with the ICCAT acronym and saying that it stood for the "International Commission to Catch All Tuna."[42] Over the years, environmentalists revised the play-on-words and labeled ICCAT the "International *Conspiracy* to Catch All Tunas." The phrase stuck and was still alive in media representations of ICCAT and in the corridors of the commission meetings I attended some twenty years later. Its force stems not only from its moralizing, satirical, confrontational style. It also stems from its ability to starkly frame ICCAT in world opinion as negligent, inept, unable or unwilling to carry out its mandate of marine conservation, broadly defined. In this sense, Safina helped to set the terms against which ICCAT had to remake itself in the twenty-first century.

The Shift to Editorialize the News

The first campaign to "save" Atlantic bluefin tuna emerged during the early 1990s and focused on the western Atlantic. The second campaign focusing on the eastern Atlantic from the mid-2000s took place in a very different journalistic context. Although both campaigns shared similar grievances, the conditions for their emergence differed in meaningful ways and shaped how the frames circulated.

Before documenting the campaign that took ICCAT member states to task after 2005, I want to acknowledge several transformations contributing to the upsurge in talk about "saving" Atlantic bluefin tuna. These include the growth, diversification, and professionalization of NGOs dedicated to the environment; the reduced investment in international news bureaus amid declining revenues and intensifying expectations for profit; the distrust in political elites amid calls for "small government" under a "neoliberal" paradigm; the rise of satellite and online news providers; and the growth of digital technologies.[43]

The 2000s were a time of wrenching transition for print newspapers aiming to bring content online. The rhythm of reporting and the pattern of media consumption changed journalism dramatically. Morning dailies on the newsstand gave way to news experienced virtually any time of day over the Internet. Michael Warner explains: "Circulation organizes time," and time organizes circulation.[44] Audiences came to rely on the more frequent, regular, continuous, sometimes unpredictable flow of discourse, just as the temporal capacity to circulate stories more frequently impacted the intensity of their flow. Under these conditions, environmentalists brought ICCAT out of the shadows in the mid-2000s.

Media coverage of ICCAT changed in kind and frequency of story. In the second bluefin tuna campaign, I found that 63 percent of all the articles in the *New York Times* about ICCAT were staff editorials, op-eds, letters, and most often Web blogs—not hard news. Prior to 2005, that figure stood at only 7 percent. The sharp increase in opinion pieces about the overexploitation of the Atlantic bluefin tuna by ICCAT member states illustrates a wider trend in environmental reporting: "The fastest-growing source for online news and analyses of environmental topics actually has not been the environmental news services but the blogosphere." To amplify an issue and "argument the news" characterized ICCAT's coverage post-2005.[45] The dominance of editorializing created a genre of news reporting that ossified the frame and stiffened the resolve to rescue this one fish. The story about the "imperiled" Atlantic bluefin tuna reemerged in the mid-2000s and peaked in 2010. This reflected the active uptake of environmental rhetoric about the urgency to "save" bluefin tuna from a "dramatic crash" that "could push the species over the brink into commercial extinction." These are the words that appeared in the *New York Times* in one of its blogs by David Jolly dated October 13, 2010.[46]

Stories about ICCAT and Atlantic bluefin tuna became increasingly commingled. At the same time, the placement of stories within the pages of the *New York Times* became more prominent. This deeply affected the rollout of the savior plot as it did ICCAT's own functioning as an RFMO. Occasionally, stories appeared in the business section, but most often they appeared in the science section, where the blogs cited throughout this chapter were originally published. Section headings served as labeling devices that narrowed the reader's field of vision. These conventions framed the discourse, channeled it to particular audiences, and shaped to whom it was directed—that is, to mainstream environmentalists concerned about how many bluefin tuna there were, not which country could legally export how many or how small she has rapidly become. Contests over commodity empires unfolding in history, based on asymmetrical power relations and on rivalries between countries with imperial ambitions, were not on the radar of the *New York Times*.

Environmental news coverage in the United States had its own rhythms of ebb and flow. Robert Cox explains: "With the rise of an ecology movement in the late 1960s, environmental news grew in coverage and reached an early peak after Earth Day during the early 1970s. After declining in the 1980s, the mainstream media's interest in the environment saw another high point in 1989, the year of the *Exxon Valdez* oil spill in Alaska." Environmental coverage declined under President Bill Clinton in the 1990s, only to resurge under President George W. Bush over concern about relaxed environmental regulations. It ebbed again post-September 11, 2001, given the domestic preoccupation with foreign-born "terrorism," and then flowed post-2006 with the coverage of the climate crisis spurred by Al Gore's film *An Inconvenient Truth* and the publication of reports by the United Nations Framework Convention on Climate Change (UNFCCC).[47] Media coverage of ICCAT mirrors this general timeline, reflecting the *New York Times* adage "All the News That's Fit to Print." This signature slogan of arguably the most authoritative news outlet in the United States can be interpreted to mean that not every frame about environmental catastrophe could possibly be made available for public review.

The Plot Thickens in the Eastern Atlantic

The second major push to protect what was becoming the icon of over-exploited, high-seas fish the world over—bluefin tuna—emerged in 2005 and gained momentum thereafter. Now the campaign focused on the other side of the Atlantic, because there the purse-seining vessels in the mid-1990s began supplying the "tuna ranches" in the Mediterranean with abandon. Although the core message remained the same— "save" bluefin tuna—new captains appeared at the helm of the second transnational campaign. Similar to Safina, Dr. Sergi Tudela of the WWF's Mediterranean Program emerged as a reliable whistleblower backed by scientific credentials. Roberto Mielgo, a former "tuna ranch" diver, provided an insider's exposé of the industry. In spring 2005, while meeting in La Plaza Mayor in Madrid, only a short metro ride from the secretariat's headquarters, they hatched a plan with representatives of WWF and Greenpeace Spain to protect bluefin tuna from the grip of state-backed, commercial interests. They made a formable team. Other NGOs, including the wealthy Pew Environment Group, joined in.[48] Some ICCAT delegates I met from France, Italy, and Spain thought that the focus on bluefin tuna in the eastern Atlantic was code for constraining the European fleets while letting the American ones off the hook. Bluefin tuna in the western Atlantic was not showing signs of recovery after decades of overexploitation either.

From 2005 onward, the press, including renowned journalist Andrew Revkin, then at the *New York Times,* picked up the bluefin story from the eastern Atlantic and ran it for several years. I found that Revkin and reporter David Jolly together accounted for 56 percent— Jolly 41 percent—of all stories in the *New York Times* about ICCAT from 2005 to 2012. These two authors were the principal interlocutors rendering ICCAT public. The stories they told resembled the usual tropes and familiar frames of the bluefin tuna campaign from the early 1990s. Prepackaged and ready to travel, these repeated story lines were the insatiable appetite for high-priced fish in East Asia; the politicization of the science; the uncontrolled, rapacious industrial fishers stealing bluefin on the high seas; the "tragedy of the commons"; and the efforts to ban the international trade of a commercial fish by listing the Atlantic bluefin as "endangered" under the CITES treaty. An editorial in the *New York Times* characterized the situation bluntly in a headline from November 17, 2007: "The Bluefin Slaughter."

Over time, the narrative gained momentum and circulated at a quickened pace, but the commission at first appeared unable or unwilling to issue robust reforms. In 2008, ICCAT member states adopted export quotas that were nearly double what the scientific advisory committee recommended. "Conservationists howled," reported Paul Greenberg in the *New York Times*. "But by the time ICCAT met again, in November 2009, environmentalists had come to home in on the historic mismanagement of Atlantic bluefin, many of them arguing that a simple reduction in catch quotas for the coming fishing season was not enough."[49] Even though the commission issued twenty-four regulations affecting the management of bluefin tuna between 2005 and 2009, environmentalists complained that the rules did not even reflect the advice of ICCAT's own scientific advisory committee. For many environmentalists, only a zero catch limit would ameliorate the overfishing crisis. Their message: halt fishing the Atlantic bluefin tuna entirely.[50]

On March 20, 2010, in response to the failed proposal to list the Atlantic bluefin as "endangered" under CITES, the editorial board of the *New York Times* matter-of-factly opined in the headline: "The Fishing Lobby Wins Again." ICCAT appeared as the lapdog of industry. The editorial portrayed the institution as incompetent, ineffective, and bought by monied interests all the way down. But this story told only a partial truth.

Eight months later, in November 2010, ICCAT met in Paris. It was the first ICCAT meeting I attended. Topping the agenda was the renegotiation of the bluefin tuna quota and its allocation. Under pressure from CITES, the head Japanese delegate on the floor asked the compliance committee, early in the meeting schedule, on November 18:

Figure 8. Greenpeace France hoisted the banner "Save Bluefin Tuna Now!" in sight of ICCAT delegates at a cocktail party along the River Seine in Paris in November 2010.

"How did you assess the catch? . . . You see that's the question posed by everyone around the world . . . [If we can't figure it out] we can't justify ourselves after this meeting." Poor data, (mis)management, and allegations of rampant illegal fishing in the news media threatened not only the future of the Atlantic bluefin tuna but the survival of the ICCAT apparatus.[51] Nearly one-hundred accredited journalists from around the world attended the opening and closing plenary sessions—by far the most I had observed in the field—raising the heat on ICCAT member states as mainstream environmentalists demanded that the bluefin be conserved responsibly.

Halfway through the meeting, the French purse-seining fleet invited all ICCAT delegates, numbering in the hundreds, to dine over wine and hors d'oeuvres at a cocktail party while touring the River Seine on a chartered vessel. Greenpeace France launched a spirited protest in the darkness of night. Activists hoisted a banner on a bridge in sight of ICCAT delegates. It read "Save Bluefin Tuna Now!"

By the end of the Paris meeting, feeling the urgency of the second transnational bluefin tuna campaign, ICCAT member states consented for the first time in recent history to heed the advice of their scientific advisory committee. They adopted a quota that they thought would allow the inventory to "rebuild." At the same time, the quota adopted would assure a level of fishing that would sustain exports, and not a level that would permit the restoration of an ocean full of big bluefin tuna fish.

For mainstream environmentalists, adhering to the advice of ICCAT's scientific advisory committee was not enough. They wanted a moratorium on catch after the failed CITES proposal, not catch levels just shy of the thresholds established in scientific recommendations. In his blog at the close of this meeting, on November 27, 2010, *New York Times* reporter David Jolly quoted the senior director of international environmental policy of the Pew Environment Group: "It is now clear that the entire management system of high-seas fisheries is flawed and inadequate."[52] Two days later, Jolly openly wondered in his blog: "Am I alone in thinking that public interest in managing endangered species justifies a little more transparency?"[53] He was not the only one frustrated by ICCAT's culture of confidentiality for the protection of trade. An environmentalist openly wept while speaking with me a few days into the conference. She felt shut out of the decision-making process, unlike her compatriots from industry who had the ear of her member state.

Four months later, on April 11, 2011, barred from entry, unable to discern happenings firsthand, the *New York Times* published the following damning words in one of its editorials: "The international body set up to conserve [bluefin tuna] has utterly failed to do its job . . . [ICCAT], responsible for regulating the bluefin catch, has for years ignored the advice of its own scientists and set quotas unsustainably high."[54] The *New York Times* did not acknowledge in the editorial that the commission actually did follow the advice of its scientific advisory committee in 2010. Once again the frame available was how big in number the bluefin tuna pie should be, not which countries were allocated how many tons of fish for export or how small the bluefin had become over the years under ICCAT's care. Outside the purview of the dominant frame was the reality that ICCAT member states had done the job asked of them by international law: to fish as hard as possible to grow a country's export economy, under the false presumption that the sea was inexhaustible.

Narrative Closure: The End of the Savior Plot

Two years later, at the 2012 commission meeting in Agadir, Morocco, the bluefin tuna quota and its allocation were up for renegotiation again. The dominant frame calling upon ICCAT member states to "Save Bluefin Tuna Now!" left no space for analyses that could lead to more critical understandings about overfishing. Instead, bourgeois readers of the *New York Times* were once again invited to enjoy an episode of the savior plot in which the expression of liberal good intentions blinded global elites from their own complicity in a predatory regime of value. The apparent end of this epic story allowed readers to feel some sense of narrative closure, relieved, unchallenged in their sensibilities. But this story surely has not ended.

ICCAT's scientific advisory committee issued the exact same advice to the commission in 2012 as it did in 2010 based on the most recent "stock assessment." The advice was the following: adopt a catch level not to exceed 13,500 metric tons for bluefin tuna in the eastern Atlantic and Mediterranean Sea and keep the level for the "stock" in the western Atlantic near 1,750 metric tons. ICCAT member states agreed to stay within the bounds of this scientific advice. Notably, the commission *raised* the quota for eastern Atlantic bluefin tuna by five hundred metric tons to 13,400 metric tons (from the agreed-upon TAC of 12,900 metric tons in 2010). In short, within the bounds of scientific advice, the revised policy *minimized* protection for the bluefin because it *raised* the quota in the East. ICCAT member states allowed more fish to be exported, not less.

In his blog post dated November 19, 2012, Jolly quoted the same representative from Pew as he had two years earlier at the end of the meeting: "[The commission] listened to the scientists this time . . . This is an organization that has had a long history of ignoring the science."[55] The tone of the Pew Environment Group had changed. Although it is true that the quota was within the bounds of what ICCAT's scientific advisory committee had recommended, the *New York Times* failed to acknowledge that ICCAT member states had already followed scientific advice two years earlier. The "generally welcomed" shift in ICCAT policy, said the *Times,* communicated to audiences that this protracted story had finally reached its end. ICCAT had changed its ways. The villain had reformed. The fish, it seemed, had been "saved."

In search of a space of purity, participants in the drama could be absolved from the harms of overfishing, immune to the structural and historical causes of managed extinction.[56] As of this writing, the *New York Times* has not published one story about ICCAT since 2012. The savior plot instead headed to the Pacific, where a new cast of characters, strikingly familiar, came to occupy the political imagination. Predictably, the bluefin was stressed there too.

Although I cannot comment with certainty on the reasons for the shift in rhetoric, it must be seen against a backdrop of changes in marine policy after 2010. I highlight two developments: one related to ICCAT rules about a new catch document program for bluefin tuna and the other related to a major change in marine policy at the level of the European Commission. Regarding the latter, more aggressive marine policy aimed at mitigating catch of commercial fish meant that the Contracting Party with the most bluefin tuna quota proportionately—the European Union—was now constrained supranationally. Under the leadership of the Greek Maria Damanaki, Commissioner of Maritime Affairs and Fisheries, the European Commission had developed its own set of rules through its Common Fisheries Policy. ICCAT member states from the European Union were bound to follow them.[57] Damanaki remarked in a statement to the *New York Times* prior to the 2010 commission meeting in Paris, "What I can say for sure is that we're going to follow scientific advice at ICCAT . . . We're not going to back down."[58] And ICCAT member states did not.

The other important policy change came in October 2011 when a report about illegal, unreported, and unregulated (IUU) fishing—funded by the Pew Environment Group—showed that ICCAT member states had exceeded the official quota for eastern Atlantic bluefin tuna by an astounding 141 percent. *New York Times* journalist David Jolly wrote in his blog on October 18, 2011: "Pew and other conservation groups have called on ICCAT member nations, which are to meet next month in Istanbul, to acknowledge the scale of the [IUU] problem and to improve the catch documentation by introducing a more rigorous electronic tracking plan." Quoting the director of (U.S.) federal fisheries policy at Pew, Jolly reported that if ICCAT member states were serious about illegal catch they would do away with paper and go electronic. They did.

During its annual meeting in Istanbul in November 2011, the IC-CAT adopted a new catch document plan for bluefin tuna that built on the program established in the first campaign from the early 1990s. ICCAT Recommendation 11–21 outlined the parameters for an electronic tracking system complete with bar codes, computers, satellites, and other technologies used to track and trace the movement of bluefin tuna in international trade. Similar to its paper forebear, the new "electronic bluefin catch document scheme" ("e-BCD") is considered an innovation in fisheries management. Atlantic bluefin tuna was once again a vector of law on the high seas.

The buzz in Istanbul about the high-tech tracking program and the sheer confidence in its worth across delegations—ICCAT member states, observers, and secretariat officials alike—spoke to a dominant sensibility: Techno-fixes will solve the overfishing problem. Questions of what new financial, administrative, and linguistic demands technology will create for the secretariat and for ICCAT member states—and whether digital documentation will be lost in cyberspace—remain open. Whatever the answers, it should be stressed that the e-BCD preserved the high-stakes bet over "red gold" for ICCAT's prestige economies. Complicating the criteria of authenticity for cross-border trade did not necessarily create a clearer path to marine conservation.[59]

Coverage of ICCAT in the second campaign heavily relied on the voice of mainstream environmentalists in the *New York Times* rather than on voices inside the fishing industry. I found that from 2005 through 2012 direct quotes came from representatives of environmental NGOs four times more often than from industry representatives. This deeply frustrated the commercial envoys I met. At lunch among a group of fisheries scientists in October 2011, one remarked that environmentalists had "defamed" the institution. He critiqued journalists for "going no further than the NGOs for a quote." I did not find officials from the secretariat quoted in the *New York Times*. (I cannot say whether requests for interviews had been made by *Times* reporters or had been refused or ignored, if requested.) The only ICCAT signatories the *New York Times* quoted were representatives from the European Union, Japan, and the United States. No commentary by government officials representing poor coastal states with an allocated bluefin tuna quota was quoted by *Times* reporters or mentioned in *Times* editorials.

Can You Bite the Hand That Feeds You?

The dominant interpretative frame made public about ICCAT positioned mainstream environmentalists as heroic figures battling to rescue an innocent creature poached on the high seas as a result of the negligent policies emanating from ICCAT. This story tells only a partial truth. It's crucial to understand just how attenuated and diminished the power of the environmentalists was during commission meetings. Behind closed doors, the influence of the environmentalists was not as prominent as it was in the public sphere. A flare-up in the compliance committee exemplifies this dynamic.

As NGOs professionalized over the years, relying on paid staff rather than on volunteers, they increasingly made use of expertise for marine advocacy. By the second bluefin tuna campaign, environmentalists attended ICCAT meetings in the main as observers (not as delegates of states). The appearance of NGOs as observers at ICCAT meetings made possible a new tactic. In 2008, ICCAT allowed observers to circulate written documents and issue oral statements on the floor while meetings were in session.[60] This development changed the contours of the "game" and altered the ways in which the stories about ICCAT's regulation of the bluefin tuna catch were spun.

In November 2012, a skirmish in ICCAT's information wars broke out at a commission meeting in the city of Agadir in Morocco. The story begins six months earlier with a press release I received via email blast from the WWF on May 31, 2012. It reported some strange activity in the Mediterranean by fleets from China and another of unknown origin spotted in the fishing grounds of bluefin tuna. It was not the only incident that the WWF (often in partnership with Greenpeace) had flagged for investigation in recent years.[61] A day later, reporter David Jolly of the *New York Times* published the story in a blog. He quoted the spokesperson for Maria Damanaki, the European Commissioner of Maritime Affairs and Fisheries: "The boats, which included 10 Chinese long-liners, were 'absolutely presumed to have been fishing illegally for bluefin,' [the spokesperson] said, adding that it was the first time Chinese vessels had been seen fishing illegally during the bluefin season." The allegation that Chinese fleets had "intrude[d] on bluefin tuna zones" came at a time when officials from the European Union and the United States were conferring on how to regulate illegal fishing while meeting in Brussels. Continuing to cite Damanaki's spokesperson, Jolly acknowledged that there was not "sufficient information to publicly accuse anyone."[62] The blog did not include comment by a Chinese official but did quote representatives from the European Union, the United States, and the Pew Environment Group. (He quoted the WWF from its press release.)

Fast-forward to day two of the ICCAT meeting in Agadir on November 13, 2012. The compliance committee convened to discuss which member states appeared to have fallen short of their treaty obligations. The documents flooding the compliance committee were so voluminous throughout the year that across delegations even the bureaucrats in charge of processing the material told me they could not read them all. Some member states were eager in committee to demonstrate their good standing in ICCAT's regulatory agreement amid this avalanche of paperwork.[63]

Japan jumped ahead of the chair's agenda to report on its own investigation of trade irregularities for the Atlantic bluefin tuna imported from Panama, also alleged by the WWF. The Japanese delegate spoke at length and, at first, at ease about the unintended, improper labeling of bluefin tuna trade documents. He emphasized how slow and burdensome the process was to investigate allegations of impropriety. Tracking down and interviewing customs agents took time. Then his tone changed quickly and became more serious. He noted that for Japan public accusations of noncompliance in the press were a concern. Oddly referencing the Chinese case published in the *New York Times*—about which Japan appeared to have no direct involvement whatsoever—the Japanese delegate asserted: "We are a little bit frustrated with [WWF's] attitude, [WWF's] action—that is, to issue a press release without making any confirmation . . . or making any contact with the flag state."

He mentioned the fleet of Chinese vessels navigating in Mediterranean waters, perplexed: "But actually when I looked up the response from China, these are fishing vessels going to West Africa for totally different fisheries." The WWF failed to issue another press release "correcting their misunderstanding." For him, the incident was, in a word, "irresponsible." Likewise, he "strongly requested" the WWF to "stop this practice." These were heavy-handed words in a diplomatic world typically accustomed to much less abrasive-sounding registers of speech.

That Japan worried about how, in its words, "to compensate for the bad image" of Contracting Parties—and of ICCAT by extension—indicates, write Chayes and Chayes, that "the fundamental instrument for maintaining compliance with treaties at an acceptable level is an iterative process of discourse among the parties, the treaty organization, and the wider public."[64] Whether "the wider public" came to know ICCAT at all and which frames dominated discourse about it are the broader questions here. Member states challenged the characterization of banditry in the press. They had to. ICCAT operates in a diplomatic world of global governance where rhetoric—not command—influences how member states make, and comply with, decisions. The credibility of the institution was on the line.

Panama and Turkey took the floor and agreed with Japan before the representative from China addressed ICCAT delegates with an air of impatience. Curt and to the point in his intervention, he thought the Japanese delegate was "correct." This came at a moment when publicly China and Japan were hotly disputing control over islands in the South China Sea. The Chinese delegate from the floor said:

> We hope that our friend from WWF will stop this business . . . WWF has a representative office in Beijing and it is easy for us to get confirmation from my bureau . . . Without such confirmation from the flag state, we believe that such irresponsible business should be stopped.

The European Union, Morocco, and Tunisia endorsed the views of Japan and China, but time ran out before the WWF could respond to them. We adjourned for an afternoon coffee break featuring Moroccan mint tea accompanied by the usual sugar cookies. The sewers were not working very well at the conference center, so to disguise the smell hotel attendants busily freshened the air with manufactured floral spray several times a day. This left a bad taste in my mouth. To escape the stench, delegates lingered outside in the Moroccan sun with smokers on break. It appeared that there was something more than just the lobby that stank.

Had the WWF been scolded for doing its job of oversight, or did the WWF overstep its bounds? How could the WWF ever confirm what ICCAT member states knew, because, like all observer delegations, it was without access to password-protected compliance documents?

The following day, the compliance committee resumed its work according to the schedule published in advance of the conference. Opening the session, the compliance committee chair—elected by the commission—immediately turned to the role of observers when submitting documents about alleged noncompliance. A member of the U.S. delegation and federal government employee, who was credited with strengthening compliance during his tenure, the chair reminded delegates: "It is not the business of this committee to discuss the communications policy of NGOs." The chair wanted to restrict discussion to substantive matters and return to open items on the agenda, including investigations of noncompliance that had yet to be resolved from previous years.

After a slight delay from technical failures in the microphone, the WWF finally took the floor and responded. It was the only oral intervention it made during the entire eight-day conference. ICCAT member states dominated proceedings. Its representative asserted in this lone intervention that the WWF had always been "scrupulous" in supplying information about possible infractions and took the necessary time to study them before going public, adding that the WWF remained available to the Contracting Parties and reserved "the right to take the floor as required by the discussion." The WWF remained "extremely loyal" to ICCAT, the representative said, and acted out of its authority to hold nation-states accountable for possible noncompliance per ICCAT Recommendation 08–09. The WWF worked within ICCAT rules to strengthen the institution, and, implicitly, asked to be heard by delegates inside meetings too, not just outside of them through the press.

After those remarks, the compliance committee chair moved on.

Days later in Agadir, with ICCAT's reputation in tatters as it struggled for credibility in the public sphere, delegates had a timely discussion about whether to revamp the commission's communications policy. The secretariat at the time was without a public-relations department, a communications officer, and an external consultant who could manage press reports. The ICCAT member states most vocally resistant to change were Japan and Morocco, with the United States (under President Barack Obama) and Brazil (under President Dilma Rousseff) eager to form exploratory committees for assessing options and cost. Without consensus, the matter was tabled for another year.

Far from the waters where bluefin tuna raced in packs at speeds rivaling a speedboat across the open ocean, ICCAT member states moved at a snail's pace. Some delegates retracted like coils and took cover inside their shells, holding close whatever "intelligence," said one, they had.

By the end of 2012, in the twilight of its career, the second campaign to "save" bluefin tuna in the eastern Atlantic appeared for readers following the drama in the *New York Times* to score several victories in the ongoing information war. The *New York Times* published reports indicating that ICCAT member states had agreed to reduce export quotas in 2010, to adopt an electronic tracking program in 2011 (to combat IUU fishing), and to follow scientific advice in 2012 (even though this had already happened in 2010). Attentive publics could be relieved of their complicity in the horror. Familiar in its missionary zeal, Judeo-Christian in its roots, the savior plot assured them that the serpent could be expelled from the Garden of Eden and the bluefin rescued from the dark forces brought to light by liberal saviors, if she had not already died for their sins. Book closed. The end. Hallelujah.

This rendition of the savior plot could not accommodate more recent developments in the ongoing chase to catch the last remnants of the giant bluefin tuna. Remarkably, the *New York Times* did not report on either of the following two stories.

In November 2017, when bluefin tuna's quota was again up for renegotiation, I calculated from afar that ICCAT member states agreed to raise the total allowable catch (TAC) for the eastern Atlantic bluefin tuna "stock" an astounding 119 percent from its lowest level in 2010 (12,900 MT) for fish caught in 2018 (28,200 MT). The TAC for the eastern "stock" skyrocketed a staggering 179 percent (to 36,000 MT) for bluefin caught in 2020. Quota for bluefin tuna in the western Atlantic increased 34 percent from 2010 to the period 2017–2020 (from 1,750 to 2,350 MT). For ICCAT decision makers, the bluefin tuna "stocks" inhabiting the Atlantic Ocean over epochs in geological time had recovered from accelerated exploitation, it seemed, in *less than a decade.* Member states dipped deeper into the inventory provided for free by Nature, worried as they were about the number of bluefin tuna in the sea for export, rather than what her shrinking size was telling them. This to me fell in line with the way fisheries management carried out its job of engineering economic growth for member states, even in a time of mass extinction.

A year later, on October 16, 2018, the European Union Agency for Law Enforcement Cooperation (EUROPOL) issued a press release with the following headline: "How the Illegal Bluefin Tuna Market Made over EUR 12 Million a Year Selling Fish in Spain." It reported: "The volume of this illegal trade is double the annual volume of the legal trade."[65] I repeat the magnitude: "double." The technology ICCAT member states used to combat IUU fishing on the high seas—the electronic bluefin catch documents (e-BCDs)—so touted as a solution by some environmentalists did not in fact "save" the bluefin from a predatory regime of value. Nowhere was it publicly noted that extracting maximized profit from monetized fish at the threshold of collapse for elite consumption was perfectly legal according to ICCAT's definition of its own mission and all the workings of its administrative apparatus.

Conclusion: Unavailable Frames in Ocean Advocacy

"In modernity," Michael Warner writes, "politics takes much of its character from the temporality of the headline, not the archive."[66] And the headline reaching the largest audience for the *New York Times*—almost 1.5 million—is the Sunday magazine[67] (Figure 9). On June 27, 2010, the foreboding banner, "Tuna's End," accompanied the subtitle: "The fate of the bluefin, the oceans and us." The journalist, Paul Greenberg, thoughtfully questioned the modern relationship with the sea and whether the bluefin would ever achieve the revered status of the mammalian whale, which most global elites do not eat today. Gripping visuals featured the bluefin as derivative, chopped up, dorsal fin here,

Figure 9. The cover of the *New York Times Sunday Magazine* dated June 27, 2010.

tail with finlets there. She appeared throughout the spread as sleek as a finely tuned automobile, her body metallic-like, her tiny scales wet as if from rain, separated from her school, alone, androgynous.

The cover story in the Sunday magazine was the only one in all forty-seven in the *New York Times* about ICCAT to mention the divide between the developed and the developing world. The source was someone located in the Global North. Buried deep into the story,

Saving the Glamour Fish

Greenberg quotes Ziro Suzuki, a Japanese scientist who once served as chair of ICCAT's scientific advisory committee:

> Developing countries firmly believe they have a right to expand their fisheries and that developed countries should reduce their fishing effort to compensate . . . In the process of trying to resolve the conflict of interest, the stocks become overfished . . . It's really just another example of the North–South problem, just like CO_2 emissions.

The thorny problem of economic (in)justice for poor coastal states—that is, former colonies and protectorates—barely got off the ground anywhere in the *New York Times*. Environmental campaigns organized to "save" bluefin tuna did not engage the harms central to parallel discourses such as those concerning climate (in)justice, or the ways in which poor countries disproportionately pay for the consequences of the climate crisis, even though they did not release most carbon into the atmosphere.

When I asked representatives of environmental NGOs about this lacuna, I got varied responses. Some dodged the question. Some turned defensive. Should they not be applauded for their efforts to reform an institution in desperate need of change?

One of the most prominent environmentalists tartly quipped, "I do not work in poverty reduction." For him, NGOs schooled in international development should have been the ones leading the charge about inequity. The complete disregard for the interdependence of all beings—that is, the compartmentalization of abstract Nature here, power asymmetries developed systematically over time through commodity empires there—was typical of contemporary technocratic thinking in ocean governance parroted by some environmentalists.

Only one environmentalist I spoke with openly acknowledged the limits of focusing campaigns on whether catch quotas were too high or whether ICCAT member states cheated the system. Both platforms could produce measurable, actionable, and therefore publishable results. To ensure their future, environmental NGOs needed to demonstrate to their constituents that their campaigns were successful. Some used their accomplishments as fund-raisings tools, if dependent on a donor base (Greenpeace, Oceana, the WWF). Others used their activism as evidence of good work for a board of directors, if endowed (Pew). Environmentalists cannot escape capitalist webs of patronage either. Michael Hardt and Antonio Negri explain the problem this way: "NGOs are completely immersed in the biopolitical context of the constitution of Empire."[68]

The more conspiratorial delegates—the ones who believed that the environmental "NGOs are fucking liars"—thought money for the second campaign post-2005 came from the United States. The United States, they claimed, wanted to stymie the economic growth of countries in the European Union by restricting their fisheries. Others thought Japan, under the thumb of the multinational corporation Mitsubishi, provided funding to the NGOs in a trick known as "astroturfing." According to this theory, an Asian powerhouse wanted to control its considerable market share in bluefin tuna by limiting supply in the face of increasing demand. This would increase the price of "red gold" and with it corporate profits. Still others thought Pew—rich from the oil giant Sunoco, based in Philadelphia—invested its endowment in ocean advocacy to get wind of developments where its own interests lay: claiming rights over the sea and its "resources" for exploitative extraction in the energy sector. Regardless of whether these rumors of dark money were true or false, they spoke to the deeper assumptions delegates held about the myriad worlds ICCAT gave expression to.

Similar to the slogans of "save" the whale, the tiger, the rain forest, the narrative of "save" the bluefin fit preexisting frames and expressed "a peculiarly bourgeois form of antimodernism." William Cronon writes of the effort to preserve "wilderness" on the American frontier: "The very men who most benefited from urban-industrial capitalism were among those who believed they must escape its debilitating effects." "Men" such as former U.S. President Theodore Roosevelt developed national parks to "preserve for themselves some remnant" of the "wild" "so that they might enjoy [its] regeneration and renewal . . . The frontier might be gone, but the frontier experience could still be had if only wilderness were preserved."[69] Giovanna DiChiro echoes this claim—rooted in imperialist nostalgia—when she writes of the mainstream: "What counts as environment is limited to issues such as wildland preservation and endangered species protection."[70] This worldview informs what counts as transnational ocean advocacy. Activists struggled to find adequate frames of reference to shed light on the various forms of predation happening on the high seas.

The mainstream variety of "blue" environmentalism emanating from the Global North has been very slow to publicize issues of environmental (in)justice, or the way the poor and disenfranchised concentrating in black, brown, indigenous communities are the most vulnerable to the pillaging of the ocean. The point is not to adopt an anthropocentric worldview by focusing on the redistribution of wealth for the world's poor, only to marginalize nonhuman natures once again. Reentrenching human exceptionalism is the least of my aims. Instead I want to point out, as DiChiro does, that many grassroots activists seeking environmental justice "explicitly undertake a critique of modernist and colonial philosophies of unlimited progress, unchecked development, the privileging of Western scientific notions of objective truth and control of nature, and the hierarchical separation between nature and human culture."[71] The dominant frames wed to the savior plot could not accommodate these critiques. Instead they pictured a privileged liberal elite haplessly stumbling upon a scene so awful that only inept, voracious, corrupt "others" could possibly be blamed for inflicting such havoc on so many creatures with so little accountability.

A few of the environmentalists I met recognized the need for a more systemic critique. But I think they understood that some frames, including ones dedicated to environmental justice, were too destabilizing to the status quo and thus were not available for use in prominent press accounts. Can—and will—publicists devise narratives, representations, and symbols adequate to capture the long-term, exponential, "slow violence" of extinction, without privileging the conventional frames fostering false hope?[72] To what extent can—and will—messaging move beyond the grammar of innocence, no longer in search of safety, of purity, of redemption, making visible the invisible, systemic causes of predation?[73] How can we better excavate and become attuned to the master narratives buried in the dominant culture we have inherited—so as to identify, resist, and neutralize their power?[74]

Perhaps readers of such news outlets as the *New York Times,* in surrendering to the drama of the savior plot, have been changed. I have reason to believe at least some people I encounter today are more aware, even if tangentially, of the rapid, catastrophic disruption now under way in the ocean than when I began this research. This is important. But armored knights on white horses brandishing swords will not "save" the bluefin, or any other creature, relieving elites from harm. A fish is not only a "thing" to be "saved," like money compounding interest in a bank account. She is a being to be respected. She is a mystery to be celebrated and revered. She is a cocreator in this life. A different mode of relating to our (tuna) kin is required to chart another course. I believe our tuna kin must be respected for them ever to be "saved." This idea awaits development in the Conclusion.

4

~~~

Alibis *for* Extermination

THE MANIPULATION *of* FISHERIES SCIENCE

"The Holy Grail"

Chapter 3 demonstrated that public condemnation of the International Commission for Conservation of Atlantic Tunas (ICCAT) was trenchant, common, and widespread. Although the commission may have appeared flawed, negligent, and inept, the public imagination continued to see fisheries science as a neutral arbiter in international marine policy making. ICCAT drew on that public confidence on the home page of its Web site, its only regular interface with ordinary citizens when I was in the field:

> *ICCAT compiles* fishery statistics from its members and from all entities fishing for these species in the Atlantic Ocean, *coordinates* research, including stock assessment, on behalf of its members, *develops* scientific-based management advice, *provides* a mechanism for Contracting Parties to agree on management measures, and *produces* relevant publications. Science underpins the management decisions made by ICCAT.[1]

Bold, confident, active verbs signified that ICCAT was busy and functional, energetic and committed to marine conservation. Curtain up, ICCAT performed its role as the steward of "these species" on the high seas and spoke from a script wrapped in the authority of international expertise: "underpin[ing]" ICCAT's "management decisions" was "science."

In ICCAT's world of ocean governance, faith in the promise of fisheries science was so pervasive as to be taken for granted. To base marine policy in fisheries science was common sense. Inside the commission, it had, said one delegate to me, the miraculous powers of "the Holy Grail."

Fisheries science is foundational to how ICCAT has operated since day one. Since its formation in the 1960s, the ICCAT convention has bound member states to scientifically ground all their decisions. Perceived as an honest broker free of national interests, fisheries science assured attentive publics that the extractive capitalists aspiring for commodity empires through the nation-state would not kill the golden goose. Fisheries science, it seemed, protected valuable fish from predatory capitalism through the assurance that it underlay and informed every aspect of ICCAT's administrative apparatus.

Yet, how to achieve sound fisheries science with which to regulate fisheries soundly has not been obvious, easy, or straightforward in the ICCAT example. Ways of knowing through fisheries science produced as many questions and undesirable outcomes as they did answers and policy prescriptions. Although constituencies across the political spectrum expected fisheries science to guide policy choice, the desire to better know commercial fish exposed the very limits of that knowledge.[2] At issue is not whether fisheries science was true or false, valid or falsifiable, successful or ineffective in its attempt to conserve marine life. Nor is the problem the failure of fisheries scientists to separate "facts" from "values," as if their boundary were axiomatic, or how "politicized" the science has become, a complaint common to ICCAT delegates. These debates distract from the more potent subject of why fisheries science appeared to be immune to, or separable from, the determining power of the forces calling it into being.

As a social scientist, I am not interested or skilled in assessing the (in)ability of fisheries scientists to "objectively and accurately understand, describe and predict the dynamics of [an] external natural reality" as it relates to the lifeworld of the bluefin.[3] I cannot comment on how well or how closely fisheries scientists captured what it meant biologically or ecologically to be bluefin. Instead I asked of fisheries science these questions: Why was its promise so easily assumed and available in the imagination of ICCAT elites? Why was its authority so taken for granted among broader publics seeking answers to the question of why all the fish were gone? Why was fisheries science treated with such deference, yet, at times, simply ignored? Of the various claims to knowledge, why did the statistically derived version dominate? On the ground, in scientific committee, who could talk? What could they talk about? Who had the authority to object? In the name of what could they object?

I learned through ethnographic research that fisheries science accomplished at least two things at once in ICCAT's regulatory zone. Materially, fisheries science rationalized the predictive power of models to inventory commercial fish on the high seas, allowing ICCAT to legally constitute and make futures markets for the member states signing the trade agreement. Ideationally, fisheries science fed the myth of objectivity so necessary to legitimate and rescue the institution from the threat of its own demise. Although some fisheries scientists may have been very committed to intellectual work, the preoccupation with and investment in fisheries science was, in reality, not pursued primarily for knowledge's sake. And while the instrumentalization of fisheries science may have been prevalent, brute self-interest was not the sole motivation operating here. Fisheries science was not just a fig leaf placed over a nakedly exploitative, industrialized process. ICCAT's scientific advisory committee was not the fox guarding the henhouse. Something more complex was going on.

Following social-scientific studies of science in policy making,[4] I want to suggest that the ritualized performance of fisheries science expressed the tacit beliefs and logics inhabiting ICCAT's "audit" culture[5] based in a managerial ethos wed to the social authority of number in the inspection of accounts. Preoccupation with mathematical modeling techniques spoke to a world firmly located in a probabilistic paradigm[6] wherein a rationalist effort to contain uncertainties appeared as the most expedient way to make marine policy.[7] The domination of modalities oriented to statistical representations formed the lifeblood of ICCAT's everyday work and moments of rule. These allowed the technocrats to level gaps in knowledge and to smooth over fuzzy interpretations, disagreements, even incompetence. Numbers—slim of form and light of prose—appeared impersonal, impartial, neutral, rational, authoritative, objective, freed of language barriers and the accompanying problems of translation.[8] Like priests centuries before them, statisticians who divined mathematical models performed how rational and enlightened they were in the secular age when state power no longer found it roots in the divine right of kings and the dark arts of superstition.[9] The promise of statistical science in its ubiquity gave the artificial impression that technocrats were in command.

The interaction between scientific and juridico-political authority in the ICCAT apparatus relied on the utilitarian use of fisheries science as a source of credibility when views and interests competed.[10] Indispensable, cardinal, compulsory, the pursuit of knowledge rooted in fisheries science was more about justifying policy decisions than it was about informing them.[11] Who had the authority to question, manipulate, and massage the (un)certainty in fisheries science was part of the "game" over who controlled the supply in "red gold" and other globally traded commodities. Fisheries science was a way for ICCAT to stage its management decisions, with statistics used as a neutral criterion in the performance of disinterested brokering.[12]

In the case of the northern cod fishery, Alan Christopher Finlayson has also discovered that science, "the erstwhile 'savior,' is not the solution but part of the [overfishing] problem."[13] The frequency with which delegates (mis)appropriated or instrumentalized fisheries science is not meant to suggest that agents were duplicitous or corrupt. ICCAT member states did not simply impose through abstract command how to organize "simple theft and unequal exchange." Michael Hardt and Antonio Negri clarify: "Rather, [apparatuses like ICCAT] directly structure and articulate territories and populations. They tend to make nation-states merely instruments to record the flows of the commodities, monies, and populations they set in motion."[14] Seen in this light, fisheries science was a tool that ICCAT delegates used to make legible aggregate "populations"—rather than individual beings—as a source of wealth because what could be assessed, measured, and rendered predictable for its optimization were detailed, statistical accounts of the whole. Biopower operates through nested interdependence.

Fisheries scientists working in committee commonly asked in ICCAT meetings: Do we know with certainty at what age, or where, the bluefin spawn? At what age the bluefin die from natural causes? Does the bluefin regularly cross the entire Atlantic Ocean? If so, how many do? These questions were hotly debated, their answers enormously consequential because they shaped the politico-economic contours of the contest for commodity empires. Fisheries scientists agreed on some things. But without consensus on new ideas, inquiries, and patterns of understanding, other matters were left to another year when, perhaps, they could be resolved. Meanwhile, at the commission, ICCAT delegates at times manipulated the uncertainties in fisheries science as a means to delay or sabotage supply flows in globally traded commodities.[15] Even so, all the fuss about the authority of fisheries science suggests that ICCAT delegates had a collective interest in mobilizing it, not only to justify and legitimate the mandate of the institution but also to perpetuate their own credibility and survival. The belief that fisheries science held the key to regulatory action kept them all coming back to the "game" every year—to settle, support, counter, test, dispute, demoralize, and undercut each other in ways that (re)entrenched ICCAT's own institutional force and viability.

I will argue in this chapter that a particular kind of fisheries science gave ICCAT's regime of value a perpetual alibi for its conduct in accelerating the extermination of oceanic life post–World War II. As the public looks for answers about the rapid slaughter of big fish, the ICCAT regime can confidently claim that it was committed, wholeheartedly, to fisheries science, to a project that could solve the hitherto unsolvable, count the hitherto uncountable, know the hitherto unknowable. ICCAT ordered and manufactured its totalizing universals; it assessed how depleted fish "stocks" were; it published its findings in highly technical reports available on its Web site. All this was perfectly acceptable to its chief constituents. Meanwhile, member states coordinated their economic development and assured the market that the supply in "red gold" would continue uninterrupted. From its institutional success on all these fronts, ICCAT became the international organization in charge of brokering the differences in ways of knowing about high-seas fish for the member states exporting them.

The remainder of this chapter ethnographically renders the process by which one lone statistic was produced by the scientists specializing in fish population dynamics. This process was mandatory by the law of treaty. Every September and October in the Standing Committee on Research and Statistics (SCRS) (Appendix C), fisheries scientists acted as "knowledge brokers" when advising the commission about how many fish to catch the coming year at a threshold preventing the collapse of an entire "stock."[16] Without fisheries science, ICCAT would have no grounds for the decisions the commission issued at the close of the annual meeting every November. At commission meetings, member states transformed the scientifically derived "maximum sustainable yield" into the policy number of "total allowable catch." This determined how big the yearly catch quota should be, from which member states allocated the wedges to countries participating in the bluefin tuna fishery. In the process of scientific knowledge production, fisheries scientists participating in the SCRS—many of them hardworking, fastidious, and well-intentioned—used their expertise to minimize their trace in promoting the prospects for industrial activity on the high seas, to harness uncertainty in the performance of objectivity, and to make legible and normalize "populations" in fish "stocks" for the commission to power over sea creatures. I conclude by exploring broader thematics raised earlier in the book about science *in* geoeconomy.

Which Science? Whose Science? Everyday Hierarchies of Power and Knowledge

All this talk about science calls for clarification, because science as a project systematically organizing, building, and verifying knowledge claims is so general a term. At play in fisheries management are the various kinds of expertise that together form fisheries science.[17] Fisheries science encompasses the specializations of oceanography, marine biology, ecology, conservation, economics, ocean resource management, and, not least, fish population dynamics. It varies from freshwater environments to saltwater seas. As an academic field of inquiry, fisheries science has developed over the years its own professional debates, journals, societies, degree-granting programs through the doctoral level, and funding opportunities. Like all intellectual traditions, it has a history that varies from country to country.[18]

The attempt to systematically measure life at sea arose during the mid-nineteenth century. A founding moment came when the government of Norway recruited the son of the pioneering marine biologist

Georg Ossian Sars to discern why cod near the Lofoten Islands fluctuated so much. In less than twenty years, Norway established a bureaucratic agency to study variation in its fisheries, and equipped it with a ship, a hatchery, and laboratories. By the turn of the twentieth century, with industrial fishing fleets now spanning the globe, other rich maritime states followed the Norwegian example. Eager to keep pace with research in neighboring Canada about the commercially important inland and coastal fish of halibut, salmon, cod, haddock, and mackerel, the United States under its commissioner of fisheries called, in 1928, for the development of a "distinct branch of scientific study, which may be termed 'fishery science.'"[19] Competitive salaries, dedicated staff, and other forms of bureaucratic investment made the field attractive to would-be scientists.

Yet in reality, far from the lab and the research trips at sea, the fisheries scientist Tim D. Smith explains: "the term 'fishery science' is used as an *administrative tool* referring to all aspects of the study of *commercially valuable* marine animals, using whatever *scientific methodologies* are appropriate."[20] Over the last century, a burgeoning administrative state provided the capital and the bureaucratic know-how to facilitate economic growth in the maritime sector.[21] An economy organized around the extraction of fish could—and should—be planned. Similar to the ecologists in the formative period of the British Empire at the turn of the twentieth century, as described by Peder Anker, fisheries scientists did not think they were part of "nature's economy."[22] In the balance, the project of building the state through the regulation of "natural-resource" markets never allowed fisheries science to mature into a field independent from political and economic interests. Although fisheries science may have had a few bad apples, its "merchants of doubt" readily emerge because of the way the field itself is structured.[23] One veteran fisheries scientist bluntly said to me in September 2012: "If you don't play by the rules, [state bureaucrats] kick you out of the laboratory." These pressures continue to shape and limit the ways fisheries science has developed.

And yet the state continues to rely on fisheries science in its aspiration for "good governance." This tension was palpable in ICCAT meetings. To summarize the paradox, Finlayson quotes the fisheries scientist Jake Rice: "It is simply good housekeeping to make [fisheries] scientists civil servants." At the same time, states "are inherently auto-cratic . . . [and] require dummy justifications for their decisions; oth-erwise people would continuously challenge the power of authority. A flock of domesticated scientists, kept sympathetic to the goals of the State through selective funding and flattery, provides such a façade."[24] Although these "either/or" scenarios summarize extreme positions—fisheries science is either a rational basis for policy making or a pre-tense for politics by other means—they also misdirect and cloud the complex lived reality about the way power and knowledge work in tan-dem.[25] They also assume that the authority of fisheries science is an enduring, universal public good untouched by a state that, in fact, culti-vated expertise over time in its concern with how to finance and sustain economic expansion.

While ICCAT delegates may have had faith in the explanatory power of fisheries science, they also knew there was a lot to learn about the marine environment. Despite the fact that the Atlantic bluefin tuna is one of the most studied fish in the world,[26] fisheries scientists regu-larly discussed in ICCAT meetings that her "population" size, migra-tory patterns, spawning habits, and other ecological and behavioral factors still elude certainty. One scientist remarked during an ICCAT working group dedicated explicitly to the study of the Atlantic bluefin tuna in September 2011: "We have found another piece of this puzzle, but we don't know how many pieces there are in the puzzle to see the whole picture." In the aspiration to see the puzzle, to capture it, to de-fine it, to contain it, to solve it, the Standing Committee on Research and Statistics (SCRS) pooled its collective specializations in fisheries science and relied on the aura of quantitative expertise to validate IC-CAT's technocratic project: to control the uncontrollable, to count the uncountable, to save the unsavable, while promoting and safeguard-ing the interests of economic growth through the "harvest" of fish on the high seas. It was as if the SCRS's sole purpose was to serve human needs, apart from the needs of the biome, as maximized short-term profits flowed asymmetrically to elites.

The emphasis on facts that can be quantified, charted, put into fig-ures, and graphed made other ways of knowing marginal in the world of ICCAT. Untenable for both fisheries scientists and fisheries managers were the rich oral histories I heard from the fishers themselves. The stories the old-timers told of the days when giant bluefin tuna traversed the ocean in packs lightning fast were not valued at all. Their narratives never rose to the status of expertise, let alone of science for ocean resource management. Instead, the oral histories of experienced anglers appeared "anecdotal," dismissed because the technocrats perceived them as too interpretive and idiosyncratic, not objective and universalizable. I share these words by an exasperated former party boat owner from New Jersey, which I heard while attending a "listening session" in January 2011 in preparation for the possible listing of the Atlantic bluefin tuna under the U.S. Endangered Species Act: "Anything that we have ever said . . . was anecdotal and didn't count. That was the word [the government] put on us. Anything that we said was anecdotal." The state needed statistical measures—not unwieldy historical testimonies—to capture what was happening at sea. Numbers appeared as the most authoritative way to make marine policy.

I also heard at this "listening session" other anglers aggravated by the uncertainty of fisheries science in policy making. On a cold winter's day in Sandy Hook, on the Jersey Shore, some forty commercial fishers and charter-boat captains gathered at Fort Hancock to share their frustrations about whether the Atlantic bluefin tuna was an endangered fish. The old artillery base, now dysfunctional and leased to commercial and nonprofit entities, still rang with shots, these ones verbal. Addressing the ambitions of the fisheries scientists, one fisher commented, "You know how hard it is to count fish? There is a little tooth fairy dimension to all this." And still another, dismayed by the uncertainty of scientific knowledge, remarked:

It doesn't seem like anyone really knows what the heck is going on . . . You read an article by whomever it might be, some doctor of something from Pew that says, you know, that the oceans are going to be devoid of life in twenty years . . . And then there are fisheries like ours where they're saying that there's no overfishing occurring, and the fish is not overfished and the stock is rebounding. There's really a tremendous divergence of data sets that people are relying on . . . Nobody can really quantify [what] anything is.

The practice of issuing policy with uncertainty in the data was, in another fisher's word, "wild." And yet the fishers' livelihoods depended on it.[27] Although the state could never fully realize the abstractions on which its authority rested, technocrats continued to invest heavily in them anyway. They were as seduced as citizens were by quantification and the evidentiary protocols of number.[28]

Remarkable, then, that more than halfway through this "listening session" one attendee asked, "Can I get an acronym definition? SCRS?" Important was not his ignorance about what SCRS meant but how far these fishers were from participating in the way ICCAT performed its legal authority over high-seas fish in the Atlantic through statistical science. When wealth is concentrated in the hands of a few, it was somehow easier—that is, more "efficient"—for ICCAT member states to work with a handful of industrial extractors than many small-scale, artisanal fishers.

As a researcher deploying social-scientific methods, I have experienced fisheries scientists wary of my own knowledge claims. One scientist said to me that he thought anthropological knowledge was, in his words, "subjective," which, in this milieu, was code for not objective. He said anthropology was too laden with "perspective." He shared that the only thing he learned in an anthropological study of fisheries in the Chesapeake Bay on the Eastern Seaboard of the United States was that fishers relied on "providence" to fix their problems. How could this analysis help him? This fisheries scientist was unable to discern that the views of life in the Chesapeake and of the single-minded pursuit of abstract, universal truth were both organized around the idea of capturing wholeness, which is an aspiration as spiritual as it is scientific. Reflectivity about knowledge claims was not an intellectual virtue cultivated among the SCRS scientists I met in the "applied science" of "tuna oceanography."[29]

Other fisheries scientists understandably asked about my own methods of inquiry. I am glad they did. While the fisheries scientists were quantifying for the commission how many Atlantic bluefin tuna there were, it seemed to them that I was qualifying them for this study. Outdoors at a plaza near the secretariat while on break during a working group dedicated to bluefin tuna in September 2011, one scientist joked over *el menú del día* among a half dozen other scientists, "We're here studying fish, and she's here studying us." In fact, what I found them studying was not fish per se but statistics about them. Fisheries scientists treated the biology of the bluefin, the environment in which she lived, and the industrial impact on her life cycle (ocean acidification, pollution, and the like) as either parameters that fed the models or variables that could not be measured and therefore could not be discussed at all.

Not all scientific knowledge—even when rising to the level of the universalizable—was equally valued at ICCAT. While marine biologists reported their findings from the field—where the fish traveled, what she ate, how fast she grew—it was the statisticians who pored over numbers using the methods of fish population dynamics that had clout, influence, and credibility, especially among the policy makers. After all, fish population dynamics developed the model that produced the all-knowing, sublime number for the commission—that is, the number designating how many tons of bluefin tuna member states could export without the "stock" collapsing. Like the farmer harvesting crops, the indicator the scientists produced was "maximum sustainable yield" (MSY). It was MSY as a number cloaked in fisheries science that the managers transformed into a number cloaked in marine policy. With the juridical formula of MSY as its guide, the commission developed its own net number called "total allowable catch" (TAC). This determined the size of the pie in quota from which member states divided and distributed.

The ICCAT treaty states unequivocally the role the SCRS is meant to play. Rule 13 of the ICCAT convention reads:

> There shall be a Standing Committee on Research and Statistics
> [SCRS] on which each member country of the Commission may
> be represented. The Committee shall develop and recommend to
> the Commission such policies and procedures in the collection,
> compilation, analysis and dissemination of fishery statistics as
> may be necessary to ensure that the Commission has available at
> all times *complete, current and equivalent statistics on fishery activi-*
> *ties* in the Convention area.[30]

Susan Silbey and Patricia Ewick explain the relationship between law and science this way: "as the law operates on the spaces and the forms within which science takes place, it contributes to the production of a distinctive content: a particular kind of science," a particular kind of scientist, a particular kind of practice followed by scientists.[31] And the particular kind of science most influential at ICCAT was the statistically oriented fish population dynamics, which forecasts the "abundance" of fish "stocks" very much the way economics predicts fluctuations in commodity futures on the Chicago Mercantile Exchange. As fisheries science laid the foundation for marine policy, the ICCAT set the parameters by which the ocean's biological assets were surveyed and measured to determine the size of the inventory in "stock."

The emphasis on fish population dynamics as the king of all fisheries science in the ICCAT apparatus reflected the advisory committee's own beliefs and shaped its sociality. The scientific meetings had their own way of being set apart from the commission. Because field scientists worked closely with fishers (scientists needed to find fish at sea to study them), they were said to develop a "human bias" or "empathy" with fishers.[32] The very nature of fieldwork could disrupt the possibility for field scientists to assert claims of permanent, objective, scientific truth. Alternatively, working the sea gave field scientists street credibility, said some fisheries scientists I met, for at least they had been on a boat, seen the bluefin alive, and touched her tiny scales with their own bare hands.

Scientists with training in fish population dynamics had their own biases too. Like bureaucrats, "quantoids" gave a great deal of agency to models when "divining the statistics," as if models came preformed by God and were not made by people.[33] If statisticians became too involved in ICCAT's day-to-day operations, ensnared in the politics of fisheries management, they exposed themselves to "charges of subjectivity and suspicion among peers. Such a judgment would be the kiss of death [not only] for any hopes of promotion" but also for the trust needed to get ordinary work done.[34]

I borrow from Douglas Whynott, who nicely summarizes the turf war in fisheries science during the 1990s. I observed a similar phenomenon two decades later in the SCRS meetings I attended:

To the field biologists who had traditionally spent a few years around the docks before doing their research, the new computer modelers seemed arrogant and stiff. The old-school scientists didn't understand the new systems and didn't know how to question them, and the new-school people seemed to have all the answers anyway, since in population dynamics it was necessary to anticipate all the questions, and cover the arguments in the way a trial lawyer prepared for a case. As a result, the whiz kids, the number crunchers, as they were called, tended to advance quickly to positions of leadership.[35]

There are two points to make here. First, at the time of my fieldwork, the two rapporteurs leading a subcommittee called the Atlantic Bluefin Tuna Working Group were both specialists in fish population dynamics. (One rapporteur represented the western Atlantic bluefin tuna "stock," and the other the eastern Atlantic bluefin tuna "stock.") Their leadership positions spoke to the social hierarchy permeating ICCAT's scientific advisory committee, which placed statistical science high on the rungs of expertise. Second, fault lines among experts demonstrated that fisheries science—however married to metrics as a way of knowing—must not be treated as a uniform, wholly integrated discipline. The scientists I met in the field treated marine conservation as a continuum with radically divergent poles on either end: the ocean was teeming with fish, the ocean was running on empty. Marine conservation among fisheries scientists was as relative as it was to fisheries managers.

For fisheries managers, statistical science was privileged because field science was expensive and yielded ambiguous data. By contrast, modeling the number of fish at sea by "population" was cost-effective and yielded specific numbers. The mathematical ways of knowing were promising: with the right data and the right formula the answer to the age-old question about how many fish there were in the sea appeared within reach.[36] Like other forms of financial modeling, the "number crunchers" at ICCAT were tasked with building an abstract representation of a real-world scenario designed to represent the performance of the "stock" in "red gold" as the greatest asset in ICCAT's portfolio.[37]

What became scientific advice for the commission often relied on who produced it and who paid for it. Unlike other regional fisheries management organizations (RFMOs), at ICCAT science-based management relied on the participation of national scientists—as opposed to in-house counsel—which delegates claimed made ICCAT's scientific advice more transparent and objective.[38] ICCAT's Contracting Parties sent their own scientists to meetings—some working for government agencies, others for universities, still others for for-profit research institutes—so that collectively they could decide through consensus what advice to give to the commission.

Even though a scientist may have registered under the delegation of, say, the United States, he did not need to be a national of the Contracting Party that sent him. At the bluefin tuna "stock assessment" meeting in 2012 held in Madrid, some fisheries scientists thought a tuna lobbyist paid for a South African national to come and finesse the numbers for the U.S. delegation. By far one of the sharpest minds in the room, the South African scientist told me that this practice was commonplace. At one meeting in a different institutional setting he entered under the flag of Canada, worked for Japan, and left under the European Union. Another scientist said to me in an interview in September 2012 that a hired gun "leans in a way that favors his client . . . He does what's contracted . . . which everybody understands." Even though a fisheries scientist could be paid by industry or by an environmental NGO to advance the interest of his client, at the same time he could not push his advocacy too far, because his own status was on the line with other top scientists in the room. It is important to note the extent to which the divisions between powerful delegations and monied interests was blurred all the way down ICCAT's decision-making process, extending from informal scientific working groups through the climactic moment when the commission issued international regulations at annual meetings.

For the most part, the fisheries scientists participating in the SCRS were delegates of states. Some fisheries scientists represented industry and civil society and came—officially—as observers. Their involvement was uneven and their attendance irregular. Most fisheries scientists registering under a Contracting Party were government employees, one of whom expressed to me that he felt "caught in the middle of the fight between the NGO-hired representatives and the industry-based representatives." Industry has been known to pay for fisheries scientists to participate in state delegations, which makes the distinction between a fisheries scientist from state and industry blurred, context dependent, not clear-cut.

At the SCRS meetings I attended, the scientists officially representing the three trading blocks of the European Union, Japan, and the United States dominated discussion. Canada occasionally added its two cents. During the Atlantic bluefin tuna "stock assessment," held in Madrid in September 2012, forty-three representatives from ICCAT member states registered. Only four of them stood for developing coastal states. Algeria and Mexico sent one delegate each, Morocco two. The rest of the delegates participating in the "stock assessment" were observers, numbering twelve (myself included), and secretariat officials, numbering six. Of the sixty-four delegates at this "stock assessment," then, only four represented a developing coastal state with an allocated bluefin tuna quota—6 percent.

The disproportionate participation of the rich trading areas was a problem recognized by some of the SCRS delegates I met. While the number of Contracting Parties to the convention has nearly doubled since the 1990s (Appendix A), the participation of scientists from all Contracting Parties has not. In other words, the member states producing the scientific advice on which the marine policy depended were the same ones producing ICCAT rules during commission meetings. (See chapter 5.) The concentration of delegates from the wealthiest Contracting Parties in the decision-making processes—in both the scientific and policy domains—revealed one of the mechanisms by which global inequities infused regulatory outcomes.

Attendance did not imply that everyone spoke and was heard. In fact, only a few scientists dominated discussion. I can count them on one hand. One scientist confided, "They're afraid to take their hands off the wheel." The ones that took the floor the most were, predictably, the quantitative analysts in fish population dynamics representing member states from the Global North. These statisticians all spoke English very well. They had to, for the only language spoken at the SCRS meetings was English. The expense of translating the techno-speak live into ICCAT's other officially spoken languages of French, Spanish, and Arabic was cost prohibitive. The favored status of English gave an edge to descendants of the British and American Empires.[39] Even so, as a native English speaker, I was exhausted at the end of each day, trying to follow nuance and innuendo amid the avalanche of technical jargon, statistics, and graphs projected in PowerPoint presentations. One scientific meeting I attended packed more than fifty papers of dizzying techno-speak in a three-day period. Language competence was a precondition for having a voice in and finding career advancement through the ICCAT apparatus. The mastery of the English language was one more example of the extent to which power asymmetries were begotten and revealed in the everyday pursuit of objective scientific knowledge in the practice of ocean governance.

Busy at Work in Committee: Naturalizing Man, Humanizing Nature

It mattered that statistical science prevailed as the field that shaped the advice ICCAT member states received from its most important advisory committee. So busy was the SCRS that it held thirteen separate meetings on three continents in 2013 alone. This does not include the additional five commission-wide meetings ICCAT held to regulate the catch of high-seas commercial fish during the same year. Four scientific committees met throughout the year: the "Species Working Groups" were dedicated to scientific debates about specific sea creatures under ICCAT's remit; the "Stock Assessments" estimated the inventory of commercially important fish every few years; the annual "SCRS Plenary" adopted the scientific advice developed in all subsidiary groups; and, not least, the "Ad Hoc Committees" addressed specific challenges, such as refining the statistics Ghana produced for tropical tunas in the Gulf of Guinea.

Similar to ICCAT meetings, all decisions in the various scientific groups were arrived at by consensus. The SCRS decision-making process aligned procedurally with such supranational scientific bodies as the Intergovernmental Panel on Climate Change (IPCC). Consensus in practice tended to render scientific advice conservative, not robust—at ICCAT, weak in protection for fish but strong in the promotion of markets. The impetus was to satisfy the least common denominator and ensure that all member states remained involved and committed to the process and its outcomes.

The scientific mandate of the SCRS was clear: while the commission produced marine policy for all member states, the SCRS produced the research and statistics on which marine policy depended. At first glance, the scientists appeared to work independently from the commission: they held separate meetings and elected their own chair. But in fact their statistics depended on what ICCAT member states submitted to the secretariat about their fishing activity, while the scope and funding of their research programs depended on what the commission deemed necessary. In other words, ICCAT shaped and bureaucratically invested in what became the advice of the SCRS in the first place.

The great impetus for knowledge production came in 2008 when the commission adopted an Atlantic-wide research program dedicated to bluefin tuna, known by the acronym of GBYP (Grand Bluefin Year Programme). With the goal of developing fishery-independent data for improved "stock assessments," the GBYP gained momentum in 2009 when member states made extra-budgetary contributions. The European Union (EU), Canada, Croatia (not then a member of the EU), Chinese Taipei (Taiwan), Japan, Libya, Morocco, Norway, Turkey, and the United States pledged more than 19 million euros to be spent over a six-year period. The EU contributed the most—a hefty 80 percent. Some environmental groups and tuna operators also provided funds or in-kind support. The program bore fruit as early as March 2010 once a dedicated staff came on board and "calls for tenders" were issued to invite proposals from third parties for contracted work. Whether some of the same scientists already participating in the SCRS won bids from, and advanced their own research agendas through, the support of GBYP I leave for others to explore. Some scientists expressed this concern to me.

Who decided how to use the money fell into the hands of the GBYP steering committee. Its members included the ICCAT executive secretary, the SCRS chair, the two rapporteurs from the Atlantic Bluefin Tuna Working Group (both specializing in fish population dynamics at the time of my research), and "an outside expert with substantial experience in similar research undertakings for other tuna RFMOs [regional fisheries management organizations]."[40] GBYP incorporated the following initiatives: aerial surveys, biological sampling, the development of historical data sets, recovering data from the ICCAT database, and tagging the bluefin at sea to better understand such things as where and at what depths she traveled with her kin. Yet, one GBYP initiative was particularly noteworthy: improving statistical modeling approaches. I heard that its budgetary allowance doubled over the life of my observations.

Of all the scientific committee meetings, the SCRS plenary may have been the most formal, but the most action-packed it was not. Every year in October at a stale hotel conference center in Madrid, national scientists and a handful of observers met to collectively adopt the recommendations emerging from the subcommittees in September. The SCRS plenary was timed for all ICCAT delegates to have at least a month to mull over scientific advice prior to the commission meeting in November. Presiding over the plenary session was the SCRS chair, who attended all scientific meetings throughout the year. He needed a handle on all the research and statistics produced by the SCRS should the commissioners have a question. "Given the uncertainties we can quantify" was the hackneyed phrase used by both SCRS chairs I met. The starched tablecloths, the national flags designating each state delegation around the negotiating table, the cocktail party celebrating the scientific achievements throughout the year, even some gray suits on opening day—these symbols marked the yearly ritualized performance of expertise bent on producing universal truths about fish "stocks" for ICCAT.

Despite the trappings of the SCRS plenary, the politics of knowledge production was most poignant during meetings of the Atlantic Bluefin Tuna Working Group and its "stock assessments." Free of national markers and other formal conventions, these meetings were held at the secretariat in Madrid and had a casual feel: short-sleeved button-down shirts, flip-flops, khakis, and jeans worn by civil servants contrasted with the tailored look of some industry representatives with their crisp collars, Ray-Bans, and cell phones abuzz in the chest pocket. Notably, industry representatives did not attend the SCRS plenary at all or in its entirety. The SCRS plenary seemed to them a rubber stamp on the scientific advice already produced by the subcommittees. Industry knew it had a voice in the working groups. Industry attended working group meetings to keep abreast of the scientific findings that could affect the rules of trade. Its representatives needed to learn about, for example, techniques to determine tonnage by counting the number of bluefin tuna caught through underwater video surveillance technology. How many bluefin tuna did the purse seiners catch and release into cages at the "tuna ranches" in the Mediterranean, once bagged and tugged long distances in early summer? Might cameras below the sea surface verify how many there were?

The knowledge elaborated in the service of the bluefin "tuna ranches" exemplified the extent to which the work of the SCRS was tailored to the commission's objectives. Simply, raising carnivorous fish for "harvest" at the "ranches" exacerbated the overfishing problem: the operators must catch mackerel, sardine, and other fish in the ocean to feed and fatten the voracious bluefin in pens. Yet the ease with which the SCRS flatly ignored the unsustainability of the industrial practice was remarkable. Rather than question the practice of "harvesting" bluefin tuna at the "ranches"—and expose what this did to deplete the ocean of fish more generally—fisheries scientists in committee developed entire research programs in support of it. They saw this as their job. Here the great elephant in the room became fully visible. Fisheries scientists were not the guardians of the sea restraining the industrial expansion of well-financed ICCAT member states. The science produced through the SCRS was grafted onto a conception of marine conservation intent on securing the inventory of fish for the growth of export markets, not to ensuring an ocean full of giant bluefin tuna and other creatures of the sea.

The purpose of the Atlantic Bluefin Tuna Working Group was twofold: update the statistics on catch provided to the secretariat by ICCAT member states and debate the best possible data to feed the models so the scenarios of "abundance" could be made calculable. In these sessions statistical science took center stage and overshadowed discussion. The SCRS was a world apart from the seafaring adventures of the marine scientists admired in *National Geographic* or in the television programs by the naturalist Sir David Attenborough. No wonder some of the most preeminent scientists studying the bluefin, including the MacArthur Fellow Barbara Block, did not attend meetings of ICCAT's scientific advisory committee. Her status and contribution as a marine scientist not specializing in fish population dynamics could only be diminished there.

To create confidence in the model devised in fish population dynamics, the working group labored to patch the leaks in a ship that seemed on the verge of sinking from uncertainty. Its dedication to statistical science acted to police the boundaries as to what qualified as objective science. Debate in the working group about whether to submit a summary report to the commission about a symposium the SCRS convened earlier in the year in Morocco on the millennia-old bluefin tuna trap fishery was exemplary. Emerging out of a recommendation made by the SCRS the previous year in support of the GBYP research program, the symposium aimed to define how the trap fishery would contribute to the future work of the SCRS. The symposium's attendees recommended the traps be listed as a site of cultural patrimony under the protection of UNESCO, which they documented in a report duly sent to the SCRS for consideration.

Yet history, culture, and socioeconomics were too much for at least some of the scientists specializing in fish population dynamics. One of them commented, "This is not very scientific . . . This is not the business of science [to endorse this recommendation]. It is not within the purview of SCRS to decide . . . As soon as you open the door to one [nonscientific issue], then why stop there?" Another fisheries scientist thought the discussion should not be about endorsing the proposal or about favoring the traps as gear but about the procedural requirement to include a report summarizing the work that the SCRS conducted during the year. Without consensus about whether to report on what seemed to the statisticians like dirt, treating it like "matter out of place," the debate evaporated.[41] This incident was indicative of the hierarchy in scientific truths. Screened out was any scientific discourse that interfered with the established common sense of a committee valuing above all else the quantifiable expertise that could be presented to its constituencies as objective.

The same scientist so intent on shutting down the debate about the symposium barked orders the next day to the only woman participating in the Atlantic Bluefin Tuna Working Group in 2011 from a member state. (The only other woman present among some thirty experts at this session represented the secretariat.) Like a pitiless taskmaster, the rapporteur, also European, asked her point blank to summarize the reports submitted over the last few days, because the working group was behind schedule and the authors, the vast majority men, neglected to do it themselves. He needed the abstracts for the executive summary, which the commissioners might read! Instead of asking for a volunteer, he summoned her to do grunt work as if she was a secretary.

The masculinist ethos was nowhere more pointedly evident to me than during the SCRS meetings. The SCRS understood fisheries science as universal and rational, not local and emotional—that is, feminine. This worldview connected fisheries science to technology and economy in ways that aligned with and promoted the well-worn, centuries-old idea of "reality as a machine rather than a living organism," one that, in the words of Carolyn Merchant, "sanctioned the domination of both nature and women."[42] The scientific method has long taken as its model "patriarchal mastery over a passive feminized nature," as Ashley Dawson and others have pointed out. This model "set the terms for subsequent notions of progress through domination of the natural world."[43] The overrepresentation of men, mostly white, from the Global North at ICCAT's scientific meetings and the lack of oral intervention by the very few women who did participate showed how gender appropriate the roles were. Whether the sex of the scientists distorted the analysis of the fisheries science—like the androcentric myth of man-the-hunter skewed for so long anthropological literature—I will leave for others to explore and decide. Situating their scientific knowledge within a frame informed by the dynamics of race, class, gender, and nation was not the concern of the SCRS.[44]

As the scientists presented their PowerPoint presentations to each other, they also documented, published, and circulated their work globally in journal articles, working papers, equations, modeling programs, maps, charts, graphs, and, not least, the reports of ICCAT's scientific meetings. The secretariat coordinated some of this activity. Interested publics with some understanding of the techno-speak came to rely on this material every year for updates on the most recent scientific developments found on the ICCAT Web site. The aspiration to resolve scientific debates called ICCAT into being and conditioned its own viability. Without a new problem to solve, an old debate to settle, a revised idea to publish each year, what else would the SCRS and the secretariat officials do—except throw their hands in the air, admit defeat in the face of the unknowable, and let the profiteering proceed without restraint? The ICCAT needed fisheries science to justify its decisions about economic growth for member states, just as fisheries science needed the ICCAT to define, develop, and fund scholarly work.

"Garbage In. Garbage Out." Harnessing Uncertainty at the "Stock Assessment"

Every September the Atlantic Bluefin Tuna Working Group meets to update, calibrate, and standardize information about how many tons of fish member states caught the previous year—as allocated—and to present new research specific to the bluefin across the various subfields of fisheries science. By contrast, "stock assessments" happened once every few years. They cost money. The ICCAT determined their frequency. The SCRS "assessed" not all creatures in the ICCAT convention area. The only ones reviewed were the commercially important fish "stocks" with statistical records.

For some experts in the SCRS, "stock assessments" happened too often. The scientists needed to accumulate data points to establish whether change in the estimated number of fish was statistically significant. Despite the short time horizon between "stock assessments," the policy makers needed assurances about whether they had struck a balance between fishing too hard and fishing hard enough to supply the market. The outcome of the bluefin tuna "stock assessment" was a highly anticipated event in the world of fisheries management. What would MSY be? That is, how many net tons of bluefin tuna would fisheries scientists recommend as reasonable for member states to catch over the next few years? It was from this figure that fisheries managers determined the TAC (total allowable catch) from which member states constructed the pie and the wedges.

The rationale for "stock assessments" was simple enough: fisheries managers must decide what to catch in the future based on approximations of past catch. How many fish are there? Is the "stock" growing in size, or is it contracting? How big is the "stock" today relative to years past? While mathematical point estimates were important to "stock assessments," over time fisheries managers began to focus their attention on probability estimates to evaluate and predict risk associated with "harvesting" in this futures market. Given the precarious complexion of an enterprise forecasting how many fish will be in "stock," the probability estimates were keen to capture (un)certainty by approximation. The fisheries scientists asked: How does the bluefin behave? How do fishers interact with her? How will the "stock" respond next year if we fish a little harder this year? What will the effect of fishing a little harder this year be in five or ten years? At what probability will the "stock" be overfished in years to come if we "harvest" at X rate today?

To compute these figures, the committee used confidence intervals, developed if–then scenarios, controlled for noise, looked for signals in the data, determined fit, ran regression analyses, checked for sensitivity in the runs, and so on. In the process, some fisheries scientists thought they only provided guidance to the fisheries managers at an arm's length. According to one manual on "stock assessments": "Fish are born, they grow, they reproduce and they die—whether from natural causes or from fishing. That's it. Modelers just use complicated (or not so complicated) math to iron out the details."[45] But my observations of the 2012 Atlantic bluefin tuna "stock assessment" found that this practice was not so cold, easy, and simple. The issue was not whether fish population dynamics achieved the right distance from the political agendas of various constituencies. It was that some fisheries scientists believed they achieved this distance at all. The model as "vast machine" designed to represent fish as mechanical, predictable, purposed for humans—that is, for elites in the Global North—helped to realize the detachment necessary to expunge the majesty of the bluefin from life in the world.[46]

The data that fed the models was of two types.[47] The first included numbers that appeared independent of the fishing industry, such as those made possible by bluefin tuna tagging programs and by larvae and aerial surveys conducted by field scientists. Fisheries scientists knew that new research could help them improve this kind of data, which explains the funding for the GBYP.

Even so, the production of scientific facts did not imply that numbers were freed from political influence or the values of their makers. To illustrate the point, I briefly turn to what fisheries scientists call "recruitment," or the number of juveniles "recruited" to enter a fishery based on the fish's life history. How big she grows and at what age she matures determines whether she is ripe for target by industry. The younger the age at which the fish matures, the more fish are catchable because the pool of available targets expands. By contrast, the older the age at which the fish matures, the fewer fish are catchable because the pool of available targets shrinks. Fisheries scientists know that they cannot rob from the future reproductive capacity of a fish "population," lest they deplete the inventory in "stock." Fish are not people. Females are more productive the older and bigger they are. The fisheries scientist Gary Sharp writes of recruitment: "forecasts are the ultimate objectives of science" and "credible fisheries recruitment forecasts are the Holy Grail of fisheries oceanography."[48]

At ICCAT, the debate about the size of bluefin tuna at maturity has been going on for decades. The bluefin is not like penaeid shrimp, for example, which is less susceptible to overfishing pressure, even in the face of enormous predation, because she grows quickly and reproduces very young. By contrast, the bluefin matures later than her tropical tuna kin, the yellowfin and the skipjack, and concentrates in areas where she spawns and feeds certain times of year, allowing fishers to easily find her.[49] Suffice it to say here that consensus on the factors impacting the bluefin's "reproductive success" based on the number of parents in her kin network has not been achieved at all. Constituencies friendly to the environment and to the long-term viability of the fishery favor what is called the "high recruitment" scenario, which aligns with the working hypothesis used in "stock assessments" across fisheries science: exploit fewer adults to allow the "stock" to replenish. More adults, more recruits. By contrast, industry-backed scientists proposed the "low recruitment" scenario when member states agreed to pull back on "harvesting" western Atlantic bluefin tuna during the 1990s, when the first organized campaign to "save" the bluefin was at its height. The "low recruitment" scenario suggests that there has been a fundamental shift in the ecology of the Atlantic, which explains why the "stock" is not as productive today.[50] Under this scenario, to restrict catch is futile, because what limits "stock" size is not the fishers but the ecological conditions enabling the bluefin to grow early in her life cycle. (Ecological factors such as ocean acidification and underwater heat waves were not part of this rationale.) The SCRS has developed highly technical work-arounds to smooth over disagreement. Consensus must be achieved. Some of the fisheries scientists I met did not even want to broach the topic about the handful of fisheries scientists who poisoned the waters and shifted the goal posts, even though this legacy is still felt today. The point: politics seeped into the biological and ecological sciences considered by the SCRS and heavily influenced the forecasting models.

In addition to fishery-independent data, other data depended on what the member states reported as catch. The ICCAT convention requires member states to submit information to the secretariat about how many fish they catch, where they catch them, by what gear they catch them, and so on. Member states are duty-bound by treaty to self-report catch so that fisheries scientists may assess "abundance." Fisheries scientists knew that fisheries-dependent data could be unreliable because member states commonly underestimated catch, similar to the way a tax evader underreports income to put a little extra in the pocket. So important is data submission to ICCAT's bureaucratic survival that at the 2011 commission meeting in Istanbul the European Union advanced and won support for the binding ICCAT Recommendation 11–15. It established that all Contracting Parties must report catch statistics—even if at zero—in accordance with data-reporting requirements. If member states did not report catch, then, says the Recommendation, they "shall be prohibited from retaining such species as of the year following the lack [of] or incomplete reporting until such data have been received by the ICCAT Secretariat." In short: no data, no fish for export. Member states could lose their right to fish their quota, even if none were caught, should they have fallen short of their reporting requirements.

Countries were required by treaty not only to submit their data but also to submit their data on time. The "stock" assessors needed to standardize data before they could plug the numbers into the models. The time pressure on state bureaucrats from capital to produce fisheries data by early September fast on the heels of the summer fishing season (and August holidays) was enormous. Missed reporting deadlines frustrated the scientists in committee because it held up their work. Not all member states, especially if poor, were successful in timely data submission. A lack of "capacity" based in underdeveloped infrastructures, language barriers, overburdened staffs, and lack of funding had the potential to impact the integrity of the modeled outcomes.

Although the models captured fishery-dependent and independent data, at the same time, there was not always sufficient data available for the fisheries scientists to advise the policy makers about what should be done. The model could not account for such disturbing trends as ocean acidification and fluctuations in warming ocean temperature. Both significantly impact the reproductive capacity of fish. Nor could the models account for the bluefin tuna exports from the nation-states not party to the ICCAT treaty. Similarly, the model could not account for illegal, unreported, and unregulated (IUU) bluefin tuna trafficked on the lucrative black market. Nor could the model capture the decreasing size of the bluefin over time, how fish interacted with one another, or what impact regulation and state subsidies had on how hard any given country fishes. The data on which the scientists depended was riddled with these and other uncertainties.

One scientist turned to me and said, "Garbage in. Garbage out."

Similar to modelers in the field of finance, the specialists in fish population dynamics know what they do not know, and know what they would like to know more about, but they "have no good way to measure what they don't know," let alone "how to measure it probabilistically."[51] Modeling "stubbornly assumes that all uncertainties about the future are quantifiable. That's why it's a model of a *possible* world rather than a theory about the one we live in."[52] Models are at once proximate to the phenomenon they seek to simulate and they operate at a distance from it.[53] As an instrument for technocratic rule, modeling simplifies, abstracts, makes impersonal its author, and generates outcomes that are precise rather than accurate. It looks forward, not backward, unable to reconstruct history and account for "shifting baselines" in the declining size of fish with each new generation.[54] Instead, it takes a present situation to predict an outcome based on a "plausible, if oversimplified, scenario" favoring not the particular or the erratic, but rather the harmonious, unified whole of a "population" in equilibrium.[55] The great simplification of the ocean into a machine with cogs of isolated fish "severely bracketed, or assumed to be constant" a complex set of relations among fish and how people interacted with them. By compartmentalizing poorly understood creatures one by one, fisheries scientists could transform each into a single element of instrumental value.[56]

Held at the secretariat's headquarters in Madrid, the 2012 bluefin tuna "stock assessment" attracted some sixty experts from across the globe. Attendance was high. The commission was keen to demonstrate the stability of the "stock" and dispel further threats to ICCAT's authority in light of the proposal to list Atlantic bluefin tuna as "endangered" under the Convention on International Trade in Endangered Species of Wild Fauna and Flora (CITES) in 2010.

Because the last "stock assessment" happened only two years earlier, the working group allowed for what one of the rapporteurs called a "strict update." The ground rule for the "assessment" was well defined: plug new numbers into the existing model and, members of the working group said, "turn the crank." They did not fine-tune, correct, rejig, or adjust the model by reconsidering new benchmarks such as how fast the fish grew, how many spawned, at what rate she naturally died. New research was off limits. One scientist remarked to the group, "Let's keep the patient alive for as long as possible, knowing that it will die soon." The model and its data appeared weak to the fisheries scientists in the SCRS this year.

The primary tasks for the 2012 "stock assessment" were focused on which data points to keep and which ones to disregard, all the while ensuring consistency in decisions to stave off human error. Similar to double-entry bookkeeping, to be precise meant that the expert had to be consistent, not accurate or exact.[57] In some instances, when data was not cut and dried, the scientists had to live with shades of gray. What is more, they had to live with a model that did not work very well. Once the scientists "turned the crank," the model produced astonishingly high numbers indicating that bluefin tuna in the eastern Atlantic and Mediterranean Sea had exploded. This was the same area where the environmentalists focused their campaigns post-2005. What was going on? Had the management measures already achieved their goal of restoring the inventory for export? Eager to capitalize on this scenario, the scientists working for industry asked, why not this outcome? A simulation can be "mathematically correct and statistically defensible" but far from the reality it forecasts.[58] The bald intrusion of capital interests irritated the statisticians who were keen to protect the sanctity of fish population dynamics. To free science from politics was easy, it seemed, but to free politics from science was another matter.[59]

While at lunch, late in the meeting schedule, scientists representing the European Union acknowledged to me that in a room full of more than sixty experts only a few of them—that is, the ones specializing in fish population dynamics—understood the complexity of the model. Sometimes tempers rose as a handful of experts bullied one another to make a point. Similar to other academic fields, fisheries science had its own rivalries, divas, and egos that needed stroking. While some fisheries scientists worried about their status, others worried about their bosses in home capitals. They had a job to keep, a promotion to score. Meanwhile, the field scientists, unfamiliar with the highly technical model, dozed off, searched the Internet for where to have dinner in Madrid that night, and chatted by text with friends over Skype. Yet, among the jealousies and disinterest, some scientists palled around, as they did during their graduate-school days. Like old buddies, they rekindled friendships or made new ones while advising the ICCAT over *vino tinto, cerveza,* and *café con leche* in the Spanish capital, sometimes late into the night at a *discoteca.*

Work continued, and continued. In committee fisheries scientists discussed the synoptic to produce knowledge narrow in scope, evident in such specialized terminology as "information plots," "catch diagnostics," "regression runs," "covariance," "jackknifing," "vectors."[60] They debated "which indices are in agreement with the catch matrix," and what color the lines in the models should be. Should they use a solid line or a dash? Some admitted that "run 2 was a misfire," and another said of another representation, "I don't see a lot of buyers."

Said one scientist to the group: they all knew that "the managers don't like to see a huge change in stock assessments over time." Critiquing a presentation that felt too idealistic for management purposes, another remarked: "I want you to have something that you can sell to a fisheries manager, and this will not sell." Not all fisheries scientists had a knack for reducing science to policy prescriptions. While the SCRS was busy "selling" its universalizing scientific truths to the commission as objective, the commission was busy selling its authority as credible to the public by using science as its alibi for the depleted number and smaller size of the once-giant bluefin tuna.

To perform objectivity, the SCRS produced products—that is, written reports—which were so laden with techno-speak that they were unintelligible to lay publics. Somewhere buried in the SCRS report was a paragraph summarizing an exceptional event marking the 2012 Atlantic bluefin tuna "stock assessment." For the first time ever, four new papers presented trade data from auction reports and customs agencies. The idea was to count the fish that went to market at the point of (whole)sale rather than to rely on the statistics that the member states submitted directly to the secretariat. Some environmentalists long concerned with illegal trade issued three of these papers. There were scientists at this meeting who believed that the Pew Environment Group in its paper misrepresented statistics to artificially inflate the figure of illegal catch as a means to endorse—and put pressure on the commission to adopt—an electronic bluefin catch document (e-BCD) program (see chapter 3).

Members of the scientific advisory committee had to agree on text to appear in extended reports and in executive summaries. The commissioners read the latter because of brevity. At one point, the group spent six minutes determining whether they had "resolved" or "analyzed" their methodologies. The latter version won. Debates over this word choice indicated how prolonged some scientific debates could become over the long term and how sanitized the scientific reports were of politics, unbeknownst to outsiders.

Day in and day out, working on Sunday and late into the night, by the end of the meeting schedule, eight days straight, one of the rapporteurs acknowledged: "We're getting tired. It's taking longer to craft words." Running out of time at the end of the meeting schedule, scrambling to finish before the secretariat's doors closed at 9 p.m., the SCRS delegates cut corners as they explained their work to the commission. The public performance of fisheries science no longer happened in a lab or on a research vessel but in a publication for the ICCAT. The report appeared removed from the authors who were the individual makers of metrics. This detachment lent the text its aura of objective truth based in fisheries science.

Of all the items addressed in a report 124 pages in length, the real concern boiled down to one: what was the magic number in the form of MSY that would assist the commission in determining the size of

Figure 10. Fisheries scientists prepared their advice in committee for the ICCAT during the bluefin tuna "stock assessment" in Madrid in September 2012.

the pie known as TAC? Because the committee agreed not to adjust the model from the previous "stock assessment," the fisheries scientists did not want to rock the boat. Their recommendation did not change from the previous "stock assessment" in 2010: maintain catch at or near recent TACs of 13,500 metric tons of bluefin tuna concentrated in the eastern Atlantic and Mediterranean Sea, and 1,750 metric tons for the bluefin tuna concentrated in the western Atlantic.

It took a great deal of work to maintain the status quo.

Sending "Signals" to the "Operators": The Normalization of Biopower

Now that the reader has a picture of the way the SCRS produced scientific advice for the commission—especially during an important "stock assessment" year—I want to briefly describe how member states made marine policy a month later when the commission convened in Agadir,

Alibis for Extermination

Morocco, in November 2012. To say that fisheries managers frustrated fisheries scientists, and vice versa, is an understatement. Some members of the SCRS were self-effacing and shunned the spotlight, uninterested in attending commission meetings. If they had, they would have experienced firsthand the way the policy makers watered down and undermined the integrity of fisheries science for political convenience. Others tried to escape commission meetings like children dodging Sunday school, and felt relief when their boss approved their absence. Still others treated commission meetings like a spectator sport and watched the politics of ocean governance from a ringside seat every year. To be fair, the work of fisheries scientists frustrated fisheries managers, for it complicated—rather than simplified—the decision-making process with all its opaque language and uncertainties.

Uncertainty in fisheries science meant something very different to fisheries management. For fisheries scientists, uncertainty informed their very analysis: what the specialists in fish population dynamics did not know they made explicit and tried to capture in models to offer the commission the best possible advice. For fisheries managers, uncertainty could indicate that they had a problem on their hands. It was they who would have to explain to attentive publics why ICCAT decided policy based on various unknowns. Tension stemmed from the way fisheries science itself was structured, for the state's bureaucratic investment in fisheries science as a check on industrial development created awkward dependencies. Frustration was also rooted in representation, or how to adequately communicate highly technical, scientific advice to laypeople.[61] At commission meetings especially—when time was short and exports dear—the job of translating highly technical fisheries science for nonspecialists was not easy for the SCRS chair.

To more simply communicate the work of the SCRS, the commission adopted ICCAT Resolution 09–12 in 2012, which required the scientists to present what is known as the "Kobe Matrix." This was used only in "stock assessments" for the commercially important fish of bluefin and bigeye tunas. Developed at a meeting of RFMOs in Kobe, Japan, in 2007, the Kobe Matrix is a graphic depiction of MSY with a twist. It plots the modeled outputs forecasting uncertainties in "stock" based on calculations of MSY—once affixed to probability estimates. The Kobe Matrix divides a two-dimensional plane into four quadrants of right angles by color: one red quadrant signifies danger or the likelihood of overfishing, two quadrants in shades of yellow mean caution, and one in green says go. When point estimates about MSY cluster in one of the quadrants, fisheries managers are supposed to adopt policy accordingly. Policy makers concentrate on the reference points crowding the red quadrant as a signal of overexploitation and the need to pull back on "harvest." The prospect of achieving an outcome absent overexploitation can be adjusted to different success rates at, say, 75 or 90 percent, once applied to different futures. What is the likelihood at a probability of 60 percent—the dominant increment ICCAT used—that overfishing Atlantic bluefin tuna would occur in five, ten, twenty years from now if *this* MSY is adopted? The lower the probability adopted, the less likely the appearance of overexploitation, and with it policy failure.

Guidelines on how to adopt the Kobe strategy occupy the bulk of this ICCAT Resolution. The aim: reduce the time fisheries managers spent reviewing lengthy, tedious files produced by the SCRS to achieve bureaucratic "efficiency," because the parameters for maximum catch were immediately known by visual graph. The Kobe Matrix collapsed futures and yesteryears into one flat, convenient plane, and reentrenched graphically the rationale that fish were first and foremost a mutable—and adjustable—inventory purposed for profitable extraction. Its geometric order echoed the grandeur of cityscapes designed by the prince in Enlightenment Europe, described by James Scott in *Seeing like a State*. The Kobe Matrix offered a view of order not from within the sea but from above or outside it, which could never be replicated no matter how intently planners aspired for a synoptic, rational vision of ocean resource management. Like a property grid in a new real-estate development, the geometric splendor of the Kobe Matrix made fisheries management convenient, "efficient," surgical, despite the fact it bore no necessary relationship with life experienced at sea by fish or fisher.[62] Although MSY inaugurated a mode of evaluating the economy in the mid-twentieth century, the Kobe Matrix represented a more recent development in the financialized forms of risk management.[63] Like traders swapping interest rates, fisheries managers on behalf of their Contracting Parties bet on the future performance of an underlying entity—a fish "stock"—with the delivery date those "harvesting" years cast in the probabilities of the Kobe Matrix.

Now that MSY was graphically rendered simple in the Kobe Matrix, it was time for the commission to decide the important number of TAC, or the net export quota to be divided among member states with a bluefin tuna allocation. The ICCAT spent a week trying to figure it out behind closed doors. Some ICCAT delegates drew attention to, and expressed concern about, the scientific uncertainty produced in the SCRS. Uncertainty could be used as a stall tactic to maintain the status quo.

The most creative use of uncertainty in fisheries science I observed came early in the meeting schedule in Agadir in November 2012. The head delegate of Japan (who also served as the ICCAT chair that year) expressed concern about "very, very severe regulations" limiting the catch of the Atlantic bluefin tuna. Japan remarked on the floor:

> The only way to solve . . . uncertainty is to increase the fishing opportunity . . . but the SCRS is asking us to stay in a very severe stage . . . I strongly hope that we can have an increase in the fishing opportunities for the entire fleet so that we can have much better data.

One way to alleviate scientific uncertainty about the bluefin was to "expand the fishing opportunities"—that is, to catch more fish. It would certainly create a bigger data set! The comment spoke to ICCAT at its most instrumental. As the utilitarian thinker Jeremy Bentham quipped long ago: "The power of the lawyer is in the uncertainty of the law." Here the power of the fisheries manager lay in the uncertainty of the fisheries science.

The decision about whether to adjust the TAC came at the very end of the meeting schedule. Statistical science was used to justify, rather than inform, policy choice. ICCAT member states agreed to keep the quota level the same for the bluefin in the western Atlantic. ICCAT member states agreed to increase the quota for the bluefin in the eastern Atlantic and Mediterranean Sea moderately from 12,900 metric tons, established in 2010, to 13,400 metric tons, so as not to exceed a TAC of 13,500 metric tons as recommended during the "stock assessment." ICCAT member states abided by the principle of science-based management, just as they had done in 2010.

The proportion of who got how much wedge stayed the same, despite protests from Turkey and Algeria, among others. The head of delegation from the European Union offered the following words on the closing day when explaining the joint proposal by the EU and Japan for the pie and the wedges in the eastern Atlantic bluefin tuna fishery:

> The principles we wanted to highlight [were] first we wanted to stay strictly within scientific advice. It's a red line, which I think if we want to keep the results of the recovery plan, then we should absolutely maintain it. So, within the scientific advice, we felt [it] was possible to send a small positive signal to the operators . . . The efforts they have made . . . bear fruit. There is recovery [of the bluefin], small still, uncertain still, but there is recovery . . . It's appropriate to send the signal to all the operators in the markets.

The "signal" to the "market" referred to a scientifically plausible increase in export quota, a reward for the fishing "operators" who reduced "harvest" in previous years. Industry had called the reduction in quota in 2010 a "sacrifice." Although the rules limiting catch appeared to limit capital accumulation, the increase in quota appeared as an olive branch, if not a carrot enticing would-be violators away from IUU fishing. Fisheries science offered policy makers reason to claim the bluefin was already "recovering" in 2012, albeit slightly, because the SCRS did not adjust MSY downward. ICCAT member states appropriated the techniques for mapping and measuring risk, and recruited fisheries scientists to act as insurance brokers concerned about managing loss in capital investments, as if actuarialists telling fortunes in "red gold" for exploitative speculation.

Conclusion: Science in Geoeconomy

Years from now, when people contemplate why giant bluefin tuna vanished from the planet, ICCAT member states may say, with full confidence, that they produced, year after year, statistical indicators of relative plenty, however fraught with uncertainty they may have been. The practice of fisheries science gave the ICCAT regime an alibi for an ocean emptying of big fish post–World War II. ICCAT member states recruited fisheries science to conserve not a creature lightning fast or a fish once regularly the size of a horse. Instead, fisheries science was tasked, by treaty, to inventory the supply of bluefin tuna to conserve the export markets of ICCAT member states. It had to. Without the modern trappings of fisheries science, the ICCAT could not justify its decisions, and thus would not be credible and able to survive.

Law and fisheries science work together for their mutual reinforcement. Silbey and Ewick clarify: "Scientific authority is in part constituted by law's deference to science's claims to discover truths that [supposedly] exist prior to and independent of human institutions." In the ICCAT example, as law deferred to fisheries science to construct its claims of authority, at the same time fisheries science actualized law's own authority to regulate.[64] Scientific and legal authority's co-construction was as much a condition as it was an outcome of new techniques of biopower. Both projects relied on the perception that technocratic elites were nonaligned, objective experts disengaged from the world around them.[65] It was as if the law, the fisheries science, the technology that the technocrats promoted, once sutured to a monetized regime of value, could evade, or transmute into abstract etherealized assets according to "stock," the reality of the destruction of the ocean while refereeing capital accumulation.

Science—like law—is not a hermetically sealed sphere operating in a vacuum external to or independent from politics and other forms of influence. This chapter instantiates the empirical reality of science not only *in* society but *in* geoeconomy. This orientation to social worlds opens up exploration into what Arturo Escobar calls "a regime of objectivism," so stubborn in its reach, here in the context of ocean governance. To be detached and objective, to appear separate from and innocent of the politics of knowledge production, denies the complex powers necessary for making fisheries science into a neutral arbiter of decision making for commodity empires rabid in extraction. This does not mean that the knowledge accrued in fisheries science does not matter or should be dismissed. Its claims to truth must be taken seriously, not to reproduce its mystifications, but to understand which modes of being and what kinds of thinking were permissible while others were being devalued and disqualified.[66]

To what extent will bluefin tuna and other sea creatures feel the effects of warming ocean temperatures, of an ocean growing ever more acidic from carbon capture? How do fish experience the tiny plastics ingested by smaller creatures that live on coral reefs now bleaching? To what extent will a slower Gulf Stream from melting ice caps reroute bluefin migrations? Will she continue to accumulate in her body mercury from the burning of coal and the mining of iron? Will the dispersants from the BP oil disaster released in 2010 affect her spawning grounds in the Gulf of Mexico or prevent her from finding a mate? Will the radiation seeping from Fukushima in the Pacific one day make its way into the Atlantic and enter her body through the global network of ocean currents? Answers to these and other questions are not known with certainty. The world as a complete and predictable whole is unattainable. But the evidence is mounting inexorably that the sixth mass extinction has begun.

Far from the sea in air-conditioned conference rooms, many of the fisheries scientists called to participate in ICCAT's decision-making process—the ones best attuned to sing her majesty—were instead alienated from whatever superqualities the bluefin embodied. They could not escape ICCAT's speculative world. Some SCRS delegates, including those specializing in fish population dynamics, acted the way financial modelers do on Wall Street when predicting market behavior by exploiting risk to maximize profit for their clients. Statistical modes of accounting required fisheries scientists to value the fish above all else as a bare number in ICCAT's regulatory zone, transforming a former ocean giant into an object of scientific knowledge inventoried for her profit potential. The awesome bluefin tuna vanished in meetings, only to be resuscitated in her commodity form.

5

~~~

The Libyan Caper

A ROGUE PLAYER WINS *the* GAME

Welcome to the ICCAT "Game"

The International Commission for the Conservation of Atlantic Tunas (ICCAT) used the advisory body known as the Scientific Committee on Research and Statistics (SCRS) and its production of abstract, seemingly objective scientific knowledge as its alibi for predation. The advice the SCRS provided to the commission appeared to make the number of fish under ICCAT's remit knowable. With the update on the "stocks" in Atlantic bluefin tuna in hand, the time had come for member states to decide marine policy. In this complex zone of international relations and statecraft, of industry and environmentalism, of conceit and deceit, hundreds of delegates from across the globe staged the annual ICCAT meeting. The chief purpose of this event was to preserve the bureaucratic machinery that member states had built and consolidated since the late 1960s.

In November 2010, ICCAT member states were under enormous pressure as they met as a commission at the height of the media blitz to "save" Atlantic bluefin tuna. Environmental NGOs came to the commission meeting in Paris, France, ready to do everything in their power to change what they considered ICCAT's anemic, inadequate, negligent marine policies. This chapter takes the reader behind the scenes, backstage, inside this high-profile event.

I learned from my very first days at this commission meeting—my first—the vernacular widely adopted by delegates to describe ICCAT's social life and regulatory culture. All the delegates, whether representing state, industry, or civil society, called this and other ICCAT meetings a "game." The prevalence of this colloquialism was shorthand for how ICCAT's work got done: a bounded, repeatable competition with fixed, predictable rules structured the field of play. Just enough room for maneuver was provided so that all contestants voluntarily committed to ICCAT's regulatory action.

The characterization of ICCAT meetings as a "game" expressed how intensely delegates were aware of the presence of winners and losers, of allies and blocs, of bluffs and daring moves. The question of why the "game" was interesting at all moves regional fisheries management beyond the formal concerns of fisheries science, law, economy, and policies of state. "Gaming" at an ICCAT meeting enacted a no less important matter in international relations: the multiple contests of status rivalries. Status rivalries took place in three registers happening simultaneously: between intergovernmental organizations, between delegations, and, not least, between individual delegates.

I am not suggesting that mathematical modeling adopted by formalistic "game theory" favored in such fields as political science and economics has much use here, because this framework overrides and dismisses the very sociality of political economy I have observed as consequential to ICCAT operations and to diplomatic life broadly. Sherry Ortner's concept of "serious games" accommodates explicitly the extent to which power and inequality pervade the games of life, which, she writes, "are quite emphatically cultural formations rather than analysts' models."[1] Rather than reduce ICCAT activities to displays of raw economic self-interest and delegates to rational economic "man"—as if the supply of "red gold" was the only thing that mattered to them—the stakes are better understood as material and economic as much as they are ideal and symbolic.

Nowhere was the "game" more pronounced than when the entire commission convened annually to decide the scope and details of marine policy for the high-seas fish. As delegates finessed statistics, scientific uncertainties, and compliance reports, on rare occasion rules of procedure became sites where at least some delegations could get an edge in the match. Rules of procedure were significant because they structured the process by which ICCAT member states decided outcomes when regulating the catch of bluefin tuna and other fish on the high seas. Consensus, not voting, was the primary mode of decision making in ICCAT's regulatory zone.

To understand the mechanisms by which ICCAT member states made decisions, this chapter develops a theme introduced in chapter 2 on the power of law in process. It focuses on consensus—and the rare derogation from it—to illuminate how member states played in the ICCAT apparatus at the height of the campaign to "save" the bluefin in November 2010. It seeks to answer the following question: how did ICCAT's predatory regime of value reproduce itself, year after year, if the result of the match was unknown to attendees, not predetermined at the start of, or during, each meeting? I begin the chapter by orienting the reader to its theoretical frame and offer thoughts on consensus as the procedural norm ascendant at ICCAT and in international law generally. The rest of the chapter is dedicated to ethnographic description of the way the Libyan delegation played with conventional rules at this one ICCAT meeting. It shows that rich and rogue delegations disproportionally influenced how ICCAT decided marine policy. Asymmetrical power relations transcended the actual process of decision making. Even in the rare derogation from the procedural norm—when consensus no longer dominated proceedings—the process by which member states developed marine policy nevertheless validated and reinscribed a predatory regime of value. The chapter concludes with thoughts on the empire of bureaucracy, or the role the ICCAT apparatus plays as a structural feature of commodity empires in finance capitalism today.

"Deep Play": Contests of Economy and State, Status and Rank

"ICCAT is the only game in town."

"You play the game by trying to figure out what the other party wants, and negotiate from there."

"To some, it's a boxing match. To me, it's a judo art."

To understand ICCAT meetings as a "game"—as these quotes from delegates in conversation attest—implies that they were at once "a contest *for* something" as much as they were a "representation *of* something."[2] The ICCAT "game" was a contest *for* oceanic control and a representation *of* intense status rivalries. As the central token of value in the global sushi economy, the bluefin excited the ICCAT "game" the most. No fish more than the bluefin cost as much at market and garnered as much time in deliberations at ICCAT meetings. No other fish received as much bureaucratic investment as the bluefin, evident in the sheer number of regulations produced on her behalf.

To appreciate the extent to which the ICCAT "game" was a representation writ small of broader geoeconomic interests and of rivalries over social prestige, this chapter takes as its framing device the famous essay by Clifford Geertz titled "Deep Play."[3] I transfer from the world of the Balinese cockfight to that of the ICCAT meeting the following insight: although the potential for material gain was great, although winning a share of quota for the planet's most expensive tuna fish carried enormous economic consequence, no less important were the symbolic rewards of esteem, honor, dignity, respect, poise, dispassion, pride, masculinity/femininity, class, country—that is, status. These were all at far greater risk in the match when others were watching at plenary or in panel discussions dedicated to the prized bluefin tuna. The potential for, and the threat to, economic growth were as relevant to fisheries management as the preservation of the ICCAT status hierarchy. The need to affirm them both, to defend them, to celebrate them, to justify them, and to just plain bask in them was a driving force behind ICCAT's diplomatic life and regulatory culture.[4]

ICCAT meetings were, par excellence, what Arjun Appadurai calls "tournaments of value": a complex periodic event in the larger commodity chain of extraction, distribution, and consumption privileging those in power by virtue of the opportunity to offer certain actors a space to display their strategic skills in attempts to divert or subvert "conventional paths for the flow of things." "[C]onsequential for the more mundane realities of power and value in ordinary life," ICCAT meetings were set apart from the usual routines of economic activity and the everyday experience of the marketplace.[5] Commission meetings were important not only because the financial stakes in the lucrative bluefin tuna trade were great. They were also opportunities for delegates to "put their money where their status is."[6] No wonder delegates filled the conference hall during deliberations about bluefin tuna in anticipation of the dramas, the bets, the strategic moves attempting to master and control the ocean through this one fish. Even delegates without an allocated bluefin tuna quota attended the match—so as not to miss the theater. Although a delegate's status may not have been radically altered by the tournament's outcome, it was nonetheless affirmed or diminished momentarily by others who as mutual acquaintances encountered each other when meetings were in session elsewhere. Some delegates crossed paths in intergovernmental organizations beyond ICCAT, including the meetings of the other regional fisheries management organizations (RFMOs), the International Whaling Commission (IWC), and, particularly important to this chapter, the Convention on International Trade in Endangered Species of Wild Fauna and Flora (CITES).

On a very few occasions delegations could test the bounds of conventional rules of procedure. This chapter describes one moment when ICCAT member states departed from their habitual course and voted on—rather than consented to—regulatory action. A focus on acts of play unfolding in a zone of ocean governance emphasizes how the unscripted, the spontaneous, the dramatic were consequential to the mundane, the predictable, the rule-bound. By unpacking the ways in which power asymmetries were begotten in the very structure of the match—or which delegations were sufficiently privileged to more freely toy with norms in procedure—I hope to add another dimension to the book's overarching concern about the ways in which the actual process of regulatory production sowed the seeds of its own contradictions.

This chapter picks up and elaborates on a theme from chapter 2 about the role of indeterminacy in international lawmaking. Rather than consider a regulatory moment when ICCAT member states regularized and secured their investment in high-seas fish, as chapter 2 did, here we see what Sally Falk Moore calls the countervailing process of "situational adjustment." She explains:

> [D]espite all the attempts to crystalize the rules, there invariably remains a certain range of maneuver, of openness, of choice, of interpretation, of alteration, of tampering, of reversing, of transforming. This is more for some people than for others, more

true of some situations than others. In many circumstances the people involved exploit the rules and indeterminacies as it suits their immediate purposes, sometimes using one resource, sometimes the other within a single situation, emphasizing the fixity of norms for one purpose, exploiting openings, adjustments, reinterpretations, and redefinition for another.[7]

This mode of law in process—characterized by the openness to maneuver and situationally adjust to a given context—aligns with and demonstrates the importance of the play element in gamelike action unfolding during ICCAT meetings.

Not all delegations were at liberty to play with rules of procedure. ICCAT meetings were not the time-space for just any delegation to do as it pleased. I found that the rich and the rogue delegations maneuvered more deftly than the poor and the peripheral ones. Poor coastal states still had reason to participate. They came to the ICCAT "game" to demonstrate their good standing in international regulatory agreements.[8] The carrot was access to key players in the global commodities trade for the promise of growth in their country's export economy. In this regulatory space, the ICCAT regime must survive to ensure an inventory in "red gold," just as "red gold" ensured the ICCAT regime a prominent place in the expanding network of ocean governance. Over time this dynamic has become "self-propelling," "self-accelerating," because "the fear of losing in the competition game, of being overtaken, left behind . . . are quite sufficient to keep the . . . game going." For select players in "liquid modernity," writes Zygmunt Bauman, the "game" was its own reward.[9]

Why Favor Consensus in International Law?

This chapter unravels the interconnections between *how* ICCAT member states made decisions by illuminating *who* decided *which* substantive matters were on the agenda. The dominant mode of decision making in international law—consensus—presumes something that feels good: a technique, ideal, custom, and affectual tie of solidarity on which the rational claims, force, and legitimacy of international law rest. I recognize that for some delegations in international fora a consensus may hold the possibility for angst and discord, as we will see. Both supranational regulations and the fisheries science on which they relied were produced through consensus at ICCAT.

Consensus implied more than the presumption that all delegations agreed. It also implied that rich and rogue delegations of state may have elected to disagree on a measure, blocking a decision altogether, or they may have diluted a measure to such a degree as to render it weak and insignificant. ICCAT is a world of descent into the least common denominator, where buy-in across membership was paramount to creating policy.[10] Decisions must be made through consensus so the entire institution gave the appearance of acting jointly, as a whole, with one voice. A big tent kept all parties active, engaged, settled under one roof, without leaving any one of them out in the rain.

By contrast, voting is a procedure of last resort in international lawmaking. This has been the trend in international fora since at least the 1960s. With the number of independent countries almost tripling since World War II—at the advent of the Great Acceleration—legal scholars regard voting majorities as "increasingly useless for lawmaking decisions" because they pose the danger of disaffection among powerful alienated minorities.[11] Since "the years of decolonization," Chen Wang writes, "developing countries obtained large voting majorities in global organizations. Western countries, traditionally playing a leading role in international affairs, can find themselves in a disadvantaged position in voting processes regarding sensitive and controversial matters."[12] Consensus in international lawmaking can be understood as a procedural effect of the postcolonial organization of world affairs.[13] The shift away from voting—toward consensus—bears on similarly situated intergovernmental organizations such as the IWC, which has long been criticized for voter buyouts. Rich countries have been known to buy votes from poor countries in exchange for promises to fund development projects at home.[14]

The importance of consensus in international lawmaking cannot be overstated. Obedience to and faith in international law relies on voluntary compliance. To voluntarily consent to and comply with rules issued supranationally reflects a conception of international order conditioning how nation-states make and carry out decisions. Within treaty relations a member state cannot be bound to rules without its consent. The parties signing the ICCAT convention have voluntarily contracted to formulate and issue recommendations, some binding on member states. That participation was, in theory, voluntary, nonhierarchical, and happened according to orderly, fixed rules in time and place, set the conditions for an institution such as ICCAT to survive. The noncoercive, nonthreatening, predictable assembly of member states deciding rules in concert brought the bureaucratic apparatus of ICCAT into being and assured it life. Even after the "game" was over at the close of each meeting, the apparatus and the community formed by it lived on.

In most cases, ICCAT's decisions were the product of intense study and debate engaged in over many years. According to the general practice of international law, should a nation-state consistently reject a measure before it becomes law, it will not be bound by it. Conversely, should a measure acquire lawlike status, it becomes obligatory upon the nation-states that have not objected to it. The ICCAT convention established the parameters for a member state to object to a binding rule agreed upon at commission meetings, which must happen in writing within a certain time period. When a member state enters an objection postmeeting, it may be signaling a future action or dissatisfaction with the way a decision was made.

All ICCAT member states had an interest in reaching consensus. Consensus was the way rules entered into force and had regulatory effect. Otherwise, delegations risked the possibility of an impasse, which could erode trust for the development of future agreement. Decisions must be made through consensus so that the entire institution could appear to be functioning smoothly—rationally, flexibly, in concert, in solidarity, by voluntary consent.[15]

The norm of unanimous agreement in the making of supranational rules raises several questions. To what extent do member states hold out for special privileges before they join others in achieving regulation, thus affecting the kinds of policies adopted?[16] Despite appearances to the contrary, to what extent do decisions made by consensus remain coercive for some parties? To what extent do regulations reached through consensus inform and (re)configure power relations on land by way of the sea? To answer these questions, I turn to an in-depth discussion of a critically important ICCAT meeting in a year when member states renegotiated the quota and its allocation for the Atlantic bluefin tuna.

Let the "Games" Begin: Derogating from Norms in Procedure

To tell the story of economy and state, of status and rank, this chapter focuses on the playful actions of various delegations, most notably Libya. In the ICCAT "game" of 2010, Libya dramatically derogated from norms in procedure by challenging the viability of consensus at this commission meeting. Here I detail a rather long but vivid scene from the 17th Special Meeting of ICCAT held in Paris. The outcome of the Paris meeting cannot be grasped without this one act.[17]

What makes this moment of rule particularly revealing, pace Erving Goffman, was its strangeness, its exceptional play, its departure from the standard etiquette governing ordinary meeting life.[18] Attention to breaches and other situations that disrupt the smooth flow of normal activity through deviant acts presents an opportunity to lay bare what was really going on. To expose the asymmetrical power relations endemic to ICCAT's decision-making process, I first describe an incident that was not supposed to happen, when interactions appeared unsuccessful, short-circuited, or upset by role confusion, breaks in frame, embarrassment, and gaffes.

In the grand ballroom of the Marriott Rive Gauche Hotel and Conference Center, billed as the largest venue of its kind in Paris, early in the meeting schedule on the afternoon of Monday, November 22, 2010, the chair opened the session with a question to the delegates assembled for matters related to bluefin tuna specifically (panel 2 in Appendix C): "What is the way we are to deal with these great stakes?" His question foreshadowed an extraordinary event in ICCAT's history at a time when the bluefin tuna "stock" in the eastern Atlantic and Mediterranean Sea was showing signs of imminent collapse. It was a time when the media campaigns by major environmental NGOs were calling for a moratorium on fishing Atlantic bluefin tuna, even after a proposal to list the fish as "endangered" (similar to pandas and rhinos) had failed eight months earlier at CITES.

The secretary-general of CITES attended the opening plenary session as a gesture of cooperation with ICCAT. For some delegates, the presence of the CITES secretary-general was not a help but a hindrance, a symbol of CITES's institutional encroachment, a warning that ICCAT might be obliged to delegate its authority over trade elsewhere should its conservation measures protecting the bluefin remain weak. ICCAT's status as the body entrusted to regulate the catch of commercial fish in the open ocean was precarious because other intergovernmental organizations were threatening to intrude upon its jurisdiction.

So important was the Paris meeting that the United States sent its highest-ranking delegate ever: Dr. Jane Lubchenco, then head of the National Oceanic and Atmospheric Administration (NOAA), part of the U.S. Department of Commerce, appointed by President Barack Obama and confirmed by the U.S. Senate in 2009. Days earlier, she opened with these words in her plenary speech, which I was told had to clear Maine Republican Senator Olympia Snowe's office at the very least: "Let's remember why we are here: to protect the fish and the fishermen and the associated businesses and communities . . . In this meeting we have choices to make. Will we choose sustainability or status quo? Will we look at the scientific advice and act with prudence and precaution, or jeopardize healthy fisheries for the future? . . . The world is watching. The world is waiting." One trade expert told me that bets had been placed in "red gold" worth five hundred million euros in 2011 alone.

On this Monday, early in the proceedings, the delegation from Libya defied ICCAT's conventional procedures. Like a player in a game of chess, the head Libyan delegate began a brilliant maneuver of "situational adjustment."[19] He exploited the indeterminacy of law in the very process of policy making by bending custom to suit his delegation's ends.

Japan spoke first in the panel dedicated to the Atlantic bluefin tuna that day. Said a delegate from a West African coastal state: "Japan controls the game . . . That's why they're always speaking here." Japan wanted its opinions known: it consumes 80 percent of the planet's bluefin tuna catch, of which 47 percent came from the ICCAT convention area while I was in the field.[20] The rich island nation expressed its support for a proposal offered a few days earlier to form a "small drafting group" to, in Japan's words, "promote a discussion on strengthening control measures as well as the observer program [for bluefin tuna]." By complying with ICCAT rules and ensuring their implementation through independent observers aboard vessels, member states participating in the fishery assured that the bluefin tuna trade was legal and the fishery sustainable, so the logic went.

The Libyan delegation addressed the panel second, once recognized by the chair: "Libya is asking for the floor, yes." The bluefin tuna fishery in Libya was said to be linked to, if not controlled by, Saif al-Islam Gaddafi (second son of the former Libyan leader Muammar Gaddafi), who, as of this writing, has been released from prison having served for alleged war crimes.[21] Recall that in November 2010 NATO's "Operation Unified Protector" in Libya had not yet happened. Not beholden to or dominated by any one powerful ICCAT member state at the time—but as a petrol state with valuable commodities in its own back pocket—Libya was a loose cannon. And it fired straight into the heart of ICCAT's decision-making process.

Libya first requested a time extension. It wanted the panel to convene for an extra fifteen minutes until 6:30 p.m. that evening. Directing the flow of the agenda, the ICCAT executive secretary denied the Libyan bid for extra time. A cocktail party was scheduled for that evening aboard a yacht on the River Seine, and to ensure that the boat left the dock in a timely fashion the buses transporting the delegates had to leave the Marriott promptly. "We must also consider rush hour," the executive secretary said.

Denied extra time, the Libyan delegation advanced its proposal nonetheless. Based on the document circulated to all delegations, including observers, titled "Last Call," Libya requested a complete moratorium—or, in its words, "a biological rest"—on fishing Atlantic bluefin tuna for two years in 2011 and 2012. The head Libyan delegate said the following from the floor:

> We are in front of all NGOs, in front of all international society, in front of all future generations . . . Most of us are fully aware from the [scientific] recommendations that we are proceeding in a very slow manner in order to rebuild and conserve [the bluefin] . . . CITES is not the place where [the bluefin] is to be managed. And that's why we took very strong action together in order

to keep the management of trade and fishing under the ICCAT umbrella. So now the ball is in the playground of ICCAT . . . We promised the whole world . . . that we would take the sacrifice . . . The Paris meeting is very important, and we have to send a very good message to the world that we do deserve the management of this stock.

And with that statement, he called for a complete halt to fishing Atlantic bluefin tuna: "We believe that in one or two years a biological rest is necessary . . . Thank you, Mr. Chair." In the spirit of diplomatic etiquette, delegates observed niceties by giving thanks and using proper address when taking and yielding the floor: "Mexico is recognized." "The distinguished delegate from Morocco." "Thank you, Madame Chair." Protocol conveyed respect and order.[22] Everyday diplomatic conventions fostered civility and mutual recognition crucial to upholding state sovereignty.

With those same diplomatic refinements, the chair next recognized the delegate from the North African member state and Libyan ally, as if he was the whole of his country: "Egypt, please." The Egyptian delegate seconded the Libyan motion.

Dead silence of nearly thirty seconds followed. That was a lifetime on the floor of an ICCAT meeting. Heads turned. Delegates whispered. The proposal to freeze fishing Atlantic bluefin tuna entirely for two years seemed absurd to most member states. But none dissented. The chair, without assurance or direction from any one delegation, awkwardly looked for consensus: "It seems there is general agreement?" Chuckles reverberated in the room. Then came spontaneous applause from the observer section in the back of the room where I sat with environmentalists and industry-backed NGOs.

Eight months earlier, in March 2010, at the CITES meeting in Doha, the Libyan delegation blocked the measure to list the Atlantic bluefin tuna as "endangered" very early in the schedule. Libya thwarted any attempt to prohibit the cross-border trade in Atlantic bluefin tuna and with it any debate about the compromised status of this fish at the CITES meeting. I heard from environmentalists who attended the CITES meeting that envelopes visibly exchanged hands. At the CITES meeting, a host of Japanese lobbyists were so aggressive, I was told, that they followed decision makers into bathrooms. Delegates were not allowed to be alone for a single moment, so intense was the pressure to ensure that tunas did not cross the red line into CITES protection. To list the Atlantic bluefin as "endangered" under CITES would have radically altered the ICCAT "game."

Especially for repeat players at ICCAT, the ones who traveled the circuits of international environmental governance and its scheduled ballet of meetings throughout the year, Libya's move came as a shock. Here was that very same delegate again, exercising his voice forcefully to put forward a proposal that on its face contradicted Libya's actions eight months earlier at CITES. In Doha, he thwarted the measure to list the Atlantic bluefin tuna as "endangered." Now, in Paris, he promoted a measure to suspend the fishery. What was going on?

The head Japanese delegate weighed in. "Good afternoon. We are talking about eastern bluefin tuna, right? Or both [eastern and western bluefin tuna]?" The Libyan proposal included both sides of the Atlantic. Japan continued: "You see, we have a similar feeling to the Libyan speakers . . . As I mentioned in my opening statement, there are two main reasons . . . to oppose [a CITES listing]": inconsistency with scientific advice and a low level of compliance. "Japan is more concerned with low level of compliance . . . If we cannot improve this situation, I think probably the Libyan idea is not so bad." The world's largest importer of bluefin tuna appeared at least superficially to support the temporary closure of the fishery.

For more than thirty-three minutes, the member states with a share of quota in Atlantic bluefin tuna expressed their views about needed "capacity reductions" (or decreasing the amount of fish caught each year by decreasing the size of a fleet and the number of vessels participating in a given fishery), about the creation of marine protected areas or sanctuaries free of fishing (once a leading proposal by environmentalists), about the importance of compliance measures to ensure ICCAT's credibility in regulating bluefin tuna's lawful trade, and, not least, about whether there was any scientific merit to the Libyan proposal. The EU, Morocco, Tunisia, the Republic of Korea, Croatia (not then a member of the EU), Turkey, China, the United States, Brazil, Canada, Norway, and Iceland voiced their positions, in that order.

The Libyan delegation then took the floor again, and insisted that

ICCAT follow the advice of its scientific advisory committee. In a mixture of Arabic and English, the Libyan delegate spoke at length about whether the probabilities currently outlined by the scientists could really predict future "stock abundance" in light of illegal, unreported, and unregulated (IUU) fishing and other activities that "pollute and contaminate our seas." The latter was an oblique reference to the BP oil disaster in the Gulf of Mexico occurring months earlier near one of the places where bluefin spawn.

Next, the Libyan delegate dropped the bomb: "I am obliged, and it is my right, and I would like to make a proposal and submit it to a vote. I want everyone to formally vote. I want everyone to see the decisions, with raised hands." Was he bluffing? Did he really want to suspend the Atlantic bluefin tuna trade for two years? And why was he calling for a vote when consensus was the norm?

The European Union and Morocco pressed the chair of ICCAT's scientific advisory committee for his views on whether a two-year moratorium would help the depleted "stock" rebuild. As the figure of scientific authority, he considered "forecast probabilities" and "the uncertainties that [the scientific committee] was able to quantify." He matter-of-factly stated: "The bigger the initial step, the less we would have to do into the future to assure rebuilding . . . The rate of rebuilding would be accelerated if you had zero catches for a biological pause of a few years." He had a way of lulling the delegates with his calm demeanor, monotone, and techno-speak, at least for those who could understand it.

The chair of the panel then asked if any delegation would like to endorse the Libyan proposal. Again, Egypt seconded the measure. The vote proceeded.

To call a vote of any kind in the context of international lawmaking was a bold move, doubly so in a year when the pie and the wedge for the central token of value—the Atlantic bluefin tuna—were on the negotiating table at ICCAT. Libya ruptured ICCAT's protocol and broke with the regular, structured flow of its proceedings, which otherwise appeared as a self-regulating system of practices, conventions, and rules that smoothly determined who was (and was not) a legitimate participant in the flow of formal talk, broadly conceived.[23] Now that norms in procedure had been transgressed—that is, by forgoing consensus—the whole play world of ICCAT unraveled. Like some kind of spoilsport who violated norms by taking advantage of them, the Libyan delegate suddenly stripped ICCAT of the illusion that decisions made by consensus were in the general interest and consensually made.[24] To some delegates, the Libyan operator was "crazy." He was monopolizing the floor and questioning ICCAT's everyday practices. Once, in a previous year, I heard that he pounded the table with his shoe—a show of disrespect—like a child in a tantrum or a court jester spinning tricks.

For nearly sixteen minutes, back and forth, round and round, the commission debated the rules of procedure in a setting where consensus overwhelmingly dominated how ICCAT member states decided marine policy. How would they vote? Draw a lot? Vote in alphabetical order? Use a show of hands? A simple majority? A two-thirds majority? The chair of the whole commission was insistent: "I think we have to be very careful on what we are doing here because I want to be 100 percent sure that we are following the rules . . . This is a very serious matter. We cannot just [vote] in a rush and ignore our rules of procedure."[25] Heads shook. Delegates sneered. Some were impatient and found concern about procedure irrelevant to the actual substance of the moratorium. The ICCAT chairperson, the ICCAT executive secretary, the panel chair, and, notably, representatives of the Japanese and Libyan delegations tried to clarify the procedure according to the text of the ICCAT convention. Around the room, gentlemen's bets emerged about how each Contracting Party in the panel would vote.

The panel chair apologized for the delay and any confusion: "Ladies and gentleman, I'm sorry for the problem as a matter of procedure . . . but it is new to us." Fractures in the ICCAT performance of rational regulatory choice making had surfaced. Anxiety was high for some, while others found the stunt by Libya comical. A rumor spread in the observer section that it was only the second time in some forty years that a member state had requested a vote by roll call. Curiously— nearly a year before the Libyan stunt—*New York Times* journalist Andrew Revkin quoted a representative from the Pew Environment Group in his blog about ICCAT dated November 14, 2009: "As a rule, decisions are made by consensus, which does not leave a lot of room for consideration of even remotely controversial proposals . . . A vote can be called for, but that has happened only twice in more than 40 years—and one of those times the vote was never taken because there was no quorum present."[26] Who was this Libyan upstart causing such a stir? Delegates from various member states huddled together, whispered, furrowed their brows, tugged their chins in earnest, strategized, and then voted.

One by one, the panel chair called on each Contracting Party to declare its vote. Yes or no? Should there be a moratorium for two years in the bluefin tuna fishery on both sides of the Atlantic or not? A key player in the drama—Algeria—had not yet arrived to the meeting and thus was not on hand to cast its vote. The vote broke down as follows, the sequence drawn by lot:

Saint Vincent and the Grenadines: No.
Syria: Abstained (not in attendance).
Tunisia: No.
Turkey: Yes.
The European Union: No.
The United States: No.
Belize: No.
Brazil: Yes.
Canada: No.
China: Yes.
Croatia: No.
Egypt: Yes.
France (Saint-Pierre and Miquelon): No.
Iceland: Abstained.
Japan: No.
Republic of Korea: No.
Libya: Yes.
Morocco: No.
Mexico: Abstained.
Norway: Abstained.

Notably, the United States joined the European Union in couching its decision of a "no" as an admonishment upon ICCAT for its lack of a "more wholesome discussion of the issue," and so, as a matter of procedure, dissented. Voting left no place for a delegation to hide.

With the roll call completed, the panel chair tallied the votes:

eleven against the proposal, five in favor, and four abstentions. The panel thus rejected the Libyan proposal. The trade in the Atlantic blue-fin tuna would continue as usual.

The Libyan delegate celebrated, clapped loudly, gesticulated to the observers in the back of the room where the environmentalists clustered, and flashed a big smile. He savored the moment as a "concocted sensation" even though his status appeared as ambivalent as ever in the sea of ICCAT delegates.[27] Outside forces working to rescue the blue-fin seemed to impinge upon normal functions, at least when member states could mobilize them, or when they threatened to do so. Only rich and rogue players were nimble enough to force an adjustment to bargaining situations in the ICCAT "game."[28] The Libyan delegate was playing the "game" of the pie to play the "game" of the wedge, like some kind of master mind-bender. His performance was no sideshow, as we learn in a moment.

The meeting adjourned for the evening. It was time to go to the cocktail party. Some environmentalists refused to attend this event because its host was a French purse-seining fleet, an alleged perpetrator of IUU fishing for bluefin tuna in the eastern Atlantic and Mediterranean Sea. Although Greenpeace France launched several protests in and around the Marriott during the meeting, the liveliest one surrounded this event. Activists deployed their self-proclaimed "TunaMobile" that night. I can still see the image of it parked along the route used by the caravan of ICCAT buses led by a police escort through the streets of Paris. One of the activists stood watch outside the vehicle with folded arms. A figurine in the shape of a bluefin tuna mounted a typical, European-styled minicar painted blue. It reminded me of a long, horizontal billboard fixed atop an everyday taxi. Graffiti-looking type stretched the length of the car and read: "Sauvez moi!!"

At the party, on several inflatable kayaks surrounding the yacht, the protesters geared up in full wet suits during this damp, cool November night. They prohibited the vessel from ever leaving its quay. To quell the protest the Paris police used the water cannon on deck and their own municipal boats to send the activists off balance and pull them from their already fragile dinghies (Figure 11). There were a few arrests. Many delegates found the stunt entertaining and snapped photos, while others were irritated that Greenpeace got in their way.

Figure 11. Greenpeace France launched a protest to disrupt a cocktail party hosted by the French bluefin tuna purse-seining fleet during the 2010 commission meeting. Paris police intervened.

They wanted a free night tour of the River Seine! A few journalists with video cameras documented the demonstration, which aired on French and international news programs. In the distance appeared a banner hanging from a bridge in the view of ICCAT delegates that read: "Save Bluefin Tuna Now!" The Eiffel Tower served as its backdrop (Figure 8).

That evening I heard from an industry representative that Libya held ten thousand tons of chilled bluefin tuna in a ship floating at sea, waiting to be released to market—as did Japan.

The Show Went On

Despite Libya's remarkable play on rules that afternoon in Paris, the everyday life of an ICCAT meeting resumed as usual. Panels, working groups, and committees met and discussed fish "stock abundance." Matters of administration, finance, and compliance with ICCAT recommendations absorbed proceedings, especially when they related to bluefin tuna.[29] There was a supranational bureaucracy to run.

Delegates fiddled on mobile phones, compared notes with colleagues in the room by text over Skype, responded to e-mails, and surfed the Internet to break the monotony, depending on where each delegate stood in the ICCAT pecking order. Low-level delegates fueled rumors and entertained these kinds of thoughts:

> Does the ICCAT secretariat monitor our Internet activity during meetings? Do they hack into our personal computers and read our files? . . . Did Prince Albert II of Monaco—the head of state who backed the CITES proposal to list the bluefin as endangered in 2010—land the fish illegally at port this summer? . . . Did IC-CAT power brokers promote former janitors to accounting positions, increasing their salary fivefold, to shut them up and cook the books?

And so it goes. These and other intrigues circulated at a dizzying rate. Gossip mattered not because it was true or false but because it expressed how delegates made sense of this diplomatic world of smoke and mirrors.

Inside the cosmopolitan hotel, the conference room was without windows and received only artificial light emanating from thick chandeliers and a string of spotlights. Together they hardened a sense of time without end—it could have been any hour of day in there—like a casino hall in Vegas. Laced atop the carpet with a geometric pattern of interlocking circles were the electrical cables for the laptops, the speakers, the microphones, the projectors. Attendees needed to review the steady stream of PowerPoint slides and to see each other on-screen when taking the floor. Half-used Evian water bottles and wrappers from breath fresheners, provided by the host country, were scattered across the tables.

Meanwhile, security guards checked for valid name cards before anyone entered the meeting area. A representative from the European Union, his name tag colored green, told me that he avoided delegates with the yellow ones. Observers, including environmentalists, wore them. Badges worn by representatives from the secretariat were colored a placid blue. Semiotics confirmed the reality that ICCAT had a status hierarchy of its own.

Like any formal affair of its kind, the organization of the main conference room communicated rank. All head delegates of all Contracting Parties sat around a table arranged in the form of a rectangle, with the remainder of Cooperating Parties, state delegates, secretariat officials, and observers in rows on the outskirts.[30] Because ICCAT is above all else a forum for member states to hammer out and refine the rules governing the global fish trade, member states had front-row seats at the match.

More specifically, on one short side of the rectangle, at the head of the room, elevated on a stage draped in burgundy cloth, were the words in English: "International Commission for the Conservation of Atlantic Tunas." It anchored the importance of the following four figures for all ten days of this meeting event: the ICCAT executive secretary, who runs the secretariat in Madrid (often with the assistant executive secretary by his side); the chair of the commission (if plenary), or the chair of a panel, committee, or working group; the rapporteur, or official note taker, nominated by a Contracting Party and approved by the commission at each meeting; and, not least, the chair of the scientific advisory committee.[31] The last symbolized the esteem for and elevation of a certain kind of expertise in the making of marine policy for the high seas. His continual presence onstage reassured delegates that regulations were rooted in the best available fisheries science. His dense presentations regularly informed them of the status of fish "stocks" under ICCAT's remit, using graphs and grids and numbers and matrices that crystallized the deep commitment to a particular type of knowledge—statistical, probabilistic, technocratic—at the expense of others. The representation of knowledge imagined through numbers, technical jargon, and other abstract, specialized frameworks became simultaneously an aid and a technique for elites to capture the world of commercial fish in order to exploit it.[32]

Around the central table, the flags of each member state distinguished one from the other and gestured to the importance of nationhood. Many Contracting Parties imagined their history as seafaring and their people as mariners, and ICCAT as the place where old duels sparred and national pride relived one more time. In these displays of resource nationalism, poor member states took offense, rightly so, when the chair did not identify them properly. "Do I see a request for the floor in the back?" "Is that Senegal? . . . No . . . Ghana, please." To request the floor, delegates raised their hand or propped up vertically the rectangular gold-plated plaque with their name, observers included. Some chairs squinted from the intensity of the spotlights when calling on a delegation to speak. They often placed a hand over their eyes like a visor to better identify a delegate in an ocean of five hundred people bathed in the glare of ballroom lights.

How nimbly one moved through the "game" was linked to how well one spoke the official ICCAT languages. When meetings were in session, attendees sat with headphones strapped to their ears so that Arabic, English, French, and Spanish could be translated live across delegations from around the world. Absent from this list were the other official languages of the United Nations: Chinese and Russian. This demonstrated the relative unimportance of those delegations in the bulk trade of ICCAT fish "stocks" at the time the convention entered into force in 1969.[33] The more languages one knew and the better one's linguistic skills, the greater one's competitive advantage to partake in this tournament of value. Fluency mattered.

The language one chose to speak on the floor was significant, and could express the gravity of a point intended for a particular delegation. Emphasize something to the Japanese and the Americans? Speak English. Show solidarity with the North Africans? Speak Arabic. Moroccans and Tunisians, for example, overwhelmingly took the floor in their colonial tongue of French even though Arabic was their official national language.

At the commission meeting in Paris, the leading transnational environmental NGOs—including Greenpeace, Oceana, the Pew Environment Group, and the WWF—gifted crystal balls made by artisans in Spain to each member state as a reminder of ICCAT's mission. To explain the gesture, Pew addressed the delegates at the opening plenary session: "It is indeed your obligation to ensure that the precautionary principle prevails. No species should have to rely on crystal ball management for its future." Given the uncertainties about the number of Atlantic bluefin tuna at sea, policy makers must take precautions.[34] Pew called for ICCAT member states to close the "tuna ranches," to suspend the industrial purse-seine fleets, and to protect the spawning grounds for bluefin tuna through the adoption of "marine protected areas" (MPAs) in the Mediterranean Sea and the Gulf of Mexico. These moments illustrated the tendency for marine advocates to talk past member states and their industry representatives. As environmentalists forcefully played the "game" of relative conservation, at the same time they overlooked or dismissed the no less important "game" of economic growth.

A few days into the meeting a few environmentalists shared with me an internal circular they had developed over the years. It humorously, lightly, playfully described "The 20 Stages of ICCAT." It read in numbered order from "anticipation" and "hope" to "extreme cynicism" and "revenge fantasies" to "apathy" and "relief." It captured for me how disorienting, volatile, frenzied the experience of an ICCAT meeting was, not just for environmentalists. Other delegates were also caught in a vortex of hyperrationality, itself an emotional state. The circular distilled the extent to which ICCAT's tournament of value unleashed "heightened flows of affect"—anger, joy, excitement, sadness, delirium, passion.[35] These emotions were otherwise not always detectable in the prudent and subtle performance of diplomacy.

Despite the formal happenings in the main conference room, it was outside this space where delegates made decisions together: over lunch or coffee or a smoke, at breakfast or lunch or dinner, beside the treadmill at the hotel's gym, at the regal gala dinner. It was, in a word, "corridor diplomacy."[36] Day in and day out, from breakfast until late at night, the real negotiations about bluefin tuna happened backstage, behind closed doors in private delegation rooms, coordinated by the powerful few, with some discussion happening front stage on the floor.

While meetings were in session, delegations offered evidence and rebuttals for and against illegal fishing, while poor coastal states complained about the high cost of participating in the bluefin tuna fishery to satisfy compliance measures. What transpired on the floor was often rehearsed elsewhere or became a stage for a delegation to spontaneously play a new card, shifting the way the meeting unfolded without the prior knowledge of others. A great many delegates were mere spectators, excluded from the decision-making process. The more removed they were from the key power brokers, the more blatant the asymmetries in ICCAT's rule-making process appeared.

Some delegates shared with me that the trading blocs of Canada, the European Union, Japan, and the United States hold what one described as a "summit" a few weeks prior to the start of the commission meeting each year to broach, or perhaps coordinate, strategies. Was the commission meeting a mere formality to recite premeditated scripts? Did these delegates share trade secrets? Did they withhold them? I cannot say. There were limits to my research beyond which I could not go—even as an accredited observer.

What made the ICCAT meeting a "game" distinct from "ordinary life" was its "secludedness," its boundedness as a defined event in time and place. As Johan Huizinga wrote long ago: "The exceptional and special position of play is most tellingly illustrated by the fact that it loves to surround itself with an air of secrecy. Even in early childhood the charm of play is enhanced by making a 'secret' out of it. This is for *us*, not for the 'others.'"[37] Behind closed doors, low-level delegates, myself included, acutely felt the distinction between "us" and "them," between the delegates with and without influence. Even so, the ground on who was considered "in" and "out" constantly shifted. Status was context dependent. It was not a binary fixed in time but a continuum reliant on where one stood in the ICCAT apparatus variable in moments and from year to year.

All ICCAT member states contributed financially to the running of the bureaucracy. The financial obligations of membership were defined in ICCAT's founding convention. Contributions varied. The more fish a member state exported, the more panels of which it was a member, the greater its per capita gross national product, the greater was its financial burden.[38] Poor coastal states in arrears were named and shamed during finance committee meetings (STACFAD in Appendix C). They were obliged to explain to fellow member states why they did not ante up. Some member states—most notably the European Union—threatened to revoke a member state's voting right if it did not pay its dues. This was a penalty made available in the rules of the ICCAT convention, but I did not once observe it applied.

Although commission meetings had their own distinct character and feel every year, they also had a structure familiar to conference-goers. Ritual acts had a conjoining effect and created bonds among delegates despite the asymmetries of power. Rituals brought about a "union . . .

Figure 12. Théâtre du Merveilleux was the venue for the 2010 gala dinner hosted by the Ministry of Agriculture, Food, Fisheries, Rural Affairs, and Regional Development of France.

or in any case an organic relation" among adversaries, opposed blocs, and unequal member states by allowing delegates to feel part of a collectivity faithful to the mandate of ICCAT.[39] Difficult "games" could be suspended temporarily among the players through ritual, if only for an evening. The annual gala dinner offered by the host country for all participants (and their partners) offered a respite from all the horse-trading in the middle of the meeting schedule. Environmentalists also attended this event.

Years earlier in Bilbao, Spain, delegates had enjoyed a gala dinner at the Guggenheim Museum designed by the famous architect Frank Gehry. In Istanbul in 2011, the dinner was held at a sultan's palace overlooking the Bosporus and included a DJ and a performance by whirling dervishes.[40] In 2010, at the request of the director for fisheries and aquaculture of France, "under the high patronage of the Minister of

Figure 13. A ringleader directed ICCAT delegates to play carnival "games" during cocktail hour at the 2010 gala dinner.

Agriculture, Food, Fisheries, Rural Affairs and Regional Development," the invitation said, delegates traveled by bus to attend the "official dinner" at the Théâtre du Merveilleux (Figure 12). There delegates played carnival games during the cocktail hour in an indoor, fantastical amusement park (Figure 13). Promoters billed the location as a "palace of illusions" where delegates encountered magical lanterns, mermen sculptures, and carousel pieces inspired by fairy tales and myths from centuries past. The boundary between real life and theatrical performance was as difficult to discern when on break as it was when meetings were in session. When the cocktail hour was over and the carnival burlesque came to a close, delegates moved into the ballroom for dinner and wine. The main course included shrimp and fillet of cod. The latter was the fish whose kin had crashed and whose Atlantic fishery had closed in Canada in the early 1990s.

Attempting to Change the Settings of the "Game"

I now want to set the stage for the coming denouement, the final and decisive act in this performance when ICCAT member states drew the pie and the wedges for bluefin tuna in 2010. On the same day as the gala dinner, Turkey and Libya formally requested a reevaluation of the criteria by which ICCAT member states decided the apportionment for bluefin tuna quota, officially called the quota allocation "scheme."

In documents issued to the panel dedicated to bluefin tuna, Turkey introduced a proposal that stated, simply: "No TAC [total allowable catch] allocation shall be made *until a consensus is reached* among the members of panel with regard to the criteria to be used for such an allocation" (emphasis added). Hours after Libya's failed proposal for a moratorium on the trade in bluefin tuna, discussed earlier, the secretariat distributed another proposal by Libya. This one requested "a re-allocation of quota in a logical and synchronized way that realizes the goals of ICCAT, in order to be equitable and fair to all." Egypt echoed the appeal in its own statement—less formal than a proposal— circulated a few days later: "It is not fair that the quota is distributed according [to] unfair standards *especially for the developing countries*" (emphasis added). Member states discussed the matter in a panel on Wednesday, November 24, 2010.

Turkey asserted on the floor that the current allocation criteria "considerably facilitate[d] a conception among parties" that these benchmarks, if left unchanged, contributed to disputes and a lack of transparency in ICCAT's decision-making process. Libya more sharply maintained that no Contracting Party far removed from the Mediterranean should get a slice of the pie at all. It considered the quota allocation a legal issue and threatened to "hire an attorney" and "take [the matter] to court" because ICCAT did not have a formal dispute mechanism at the time. In the Libyan view, the criteria in use in 2010 were in violation of the United Nations Convention on the Law of the Sea (UNCLOS III), which upheld the right of each nation-state to control marine "resources," including fish, within two hundred nautical miles from its coast by virtue of an exclusive economic zone (EEZ). Libya had again challenged ICCAT's operating common sense by characterizing its traditional allocation "scheme" as unlawful by international standards.

Egypt then contended that no Contracting Party that had devastated the "stock" should be awarded a quota. This was an implicit rebuke of the fleets of the European Union that had been widely criticized by environmentalists for overexploiting bluefin tuna in the Mediterranean Sea in recent years. Egypt went on to assert that the inspectors, the reports, the production of the national fishing plans, the training activities, the ICCAT fees, the travel to meetings were all very expensive, especially for poor countries. The outlay of cash to legally participate in the fishery was disproportionate to the return on its investment, given the small percentage of bluefin tuna quota Egypt was allotted. The Republic of Korea weighed in and agreed the allocation criteria should be revisited.

It is significant that countries can air their grievances in settings such as ICCAT. But airing grievances does not necessarily mean they will be acted upon. The European Union took the floor and staunchly countered these positions. It effectively blocked consensus with one move. It suggested that ICCAT member states had done "a good job" in redefining the criteria a decade earlier and therefore the panel did not need to reopen discussion on that topic again. No other delegation intervened. Without consensus across ICCAT member states party to this panel, the debate simply evaporated. The issue was closed, at least for this year, owing to the European Union's lone, active dissent. Change by diplomacy came slowly.[41] Or at least it did not come fast enough for poor countries and the survival of a tuna once called "giant."

Settling Scores in the Paris Match

The secretariat was supposed to distribute recommendations on the Wednesday prior to the closing plenary on Saturday—to give ample time to translate documents into English, French, and Spanish and to allow each delegation adequate time to review text. But none of the delegates attending Wednesday's, Thursday's, and Friday's sessions had received any notice in their "pigeonholes" indicating that a decision was near about the size of the pie and its corresponding wedges for the Atlantic bluefin tuna. This is not to say the meeting was without paper. The secretariat distributed more than six hundred thousand photocopies at the 2010 Paris commission meeting alone.[42] Nayanika

Mathur explains: "[A]n explosion in paperwork and an ever-expanding reliance on documents" constituted "concrete evidence" of the labor expended for the "production of 'result.'"[43] The more ICCAT matured, the longer its paper trail in the performance of outcomes, solutions, and results.

The meeting schedule shifted constantly to accommodate the failure of member states to offer timely recommendations. Without proposals, there was nothing to debate on the floor and therefore nothing to consent to. As the days passed, the panels, the working groups, the committees concluded their agendas. While high-level officials conducted meetings behind closed doors, low-level delegates lingered in the halls of the Marriott, offering predictions of what would transpire next, jangling their nerves with espressos, waiting for the smoke signals, trading confidences, sharing whispered intrigues, painfully aware they had been shut out of the decision-making process.

Not until the closing day—Saturday, November 27—did the secretariat release the proposal for catch quotas in the eastern Atlantic and Mediterranean Sea at the wee hour of 2:17 a.m.[44] The European Union and Japan were its cosponsors. For bluefin tuna in the western Atlantic, the proposal by the United States and Japan arrived even later, at 4:07 p.m. on Saturday afternoon, just hours before the scheduled departure of some delegates from Charles de Gaulle airport. There were only three delegations submitting the recommendations on which the whole of the commission was to decide the future of its hottest "stock": the European Union, Japan, and the United States.[45] Many delegations were furious. The sentiment was that a few powerful players brokered deals in private, and now that so much time had been taken from the meeting schedule there was little opportunity for debate in plenary. The chance for negotiation was over.

At the close of the meeting, member states settled scores. For both the western and eastern "stocks" of Atlantic bluefin tuna, the pies proposed were just short of what the scientific advisory committee had recommended—a considerable departure from the norm in recent years. The United States presented its joint proposal with Japan: a "total allowable catch" (TAC) of 1,750 metric tons, just 50 metric tons shy of the level recommended not to exceed 1800 metric tons. Allocated shares in the West remained the same. For the eastern "stock," the Japanese delegation presented its joint proposal with the European Union: a TAC of 12,900 metric tons, or 600 metric tons less than the level recommended not to exceed 13,500 metric tons. The commission demonstrated to environmentally attentive publics that it was following the advice of its scientific advisory committee. A high-ranking delegate of state close to the negotiations remarked to me, "From now on, it's another kind of game."

In the eastern Atlantic, the quota allocations proposed were striking. Appendix B shows that Libya nearly doubled its quota, a remarkable and unusual achievement, courtesy of Algeria. Because the Algerian delegation was delayed in arriving—which I heard was owing to overlap in the timing of the conference and the observance of the Festival of the Sacrifice (Eid al-Adha)—Algeria was not on hand during negotiations to protest its loss in share. Even so, why take from Algeria if Albania and Syria did not attend the meeting at all? Shares allocated to Albania and Syria went undisturbed in the 2010 allocation. Was Algeria's loss in quota a casualty of the "abrasive relations" between Paris and Algiers under former French President Nicolas Sarkozy?[46] Did France cut deals to rob from Peter to pay Paul? I can only speculate.

All allocations for all other delegations fishing for bluefin tuna in the eastern Atlantic remained the same, although Turkey made slight gains from the Algerian loss. By coincidence, I ran into a high-ranking delegate from the European Union on the Paris metro after the close of the meeting who told me that he knew he "screwed" the Algerians but also recognized "early on" in the meeting that he had to "deal" with the Libyans. Libya's forceful strategy days earlier had paid off. Libya won the "game" of economic growth by playing the "game" of relative conservation, accomplished formally by manipulating the rules of procedure in a regulatory process open to "situational adjustment."

Presenting the proposal for the eastern Atlantic, the Japanese delegation—on behalf of its cosponsor, the European Union—detailed its provisions, including increased compliance measures and sanctions against countries such as France that had exceeded its quota in recent years. The head Japanese delegate sidestepped the issue of the skewed quota allocation altogether. His words on the floor: "Because we are in a very low [total allowable catch] level, it is not the time to realize the fishing opportunity for the new participants to fulfill their aspiration. In [the] future, if we are successful for the recovery of this stock, then . . . we will adjust the allocation keys." Under this proposal, the commission appeared unwilling to more equitably allocate quota to poor coastal states as the new entrants to the ICCAT "game." Libya was the exception this year. Some delegates told me at later commission meetings that they thought Japan rewarded Libya for its actions in CITES when Libya torpedoed the measure to list Atlantic bluefin tuna as "endangered." Negotiations created strange bedfellows when commodity empires were under threat.

In seconding its own proposal, the head delegate from the European Union took the floor, thanked Japan for its presentation, and remarked to his colleagues around the table in English, not his native French: "We hope that this proposal, if adopted by the [Contracting Parties], will open a new era of cooperation in the Mediterranean, in particular in the field of inspections, in the field of aerial surveys of a scientific nature, in the field of exchange of data between the coastal states of the Mediterranean." He then entreated other delegations around the room to, in his words, "play this game with us." Some delegates remarked to me earlier in the week that they were impressed by his English. They said that in these forums the French have a residual accent, so they took his language competence as a sign that Brussels had sent a high-ranking official to ICCAT during this important year of bluefin tuna quota negotiations. Although his performance may have enhanced his status among some delegates, others within the EU delegation expressed to me the opinion that he did not fully appreciate how technical ICCAT's work was. To my knowledge, this was his last ICCAT meeting.

In response to the proposals, various member states, including Algeria, Canada, China, Mexico, Morocco, Turkey, and Tunisia, took the floor and expressed what they called their "disappointment" in the decision-making process. Would they inhibit the commission from issuing a decision this late in the "game"? With the press and its cameras awaiting them outside the plenary, delegates knew that some decision was better than no decision at all.

The Norwegian delegation issued the most blistering remarks:

This year, as well as in the preceding years, we have witnessed a total lack of transparency in the decision-making process. There

[have] been no real discussions in panel . . . and all the negotiations took place behind closed doors with a majority of the
members of the panel on the outside. For us, this is totally unac-
ceptable. We have now before us a Recommendation that I must
admit I have not had the time to read.[47]

In a letter to the commission months later, Norway deposited a formal objection to the secretariat, again reiterating, in its words, its concern about a "lack of transparency in the decision-making process. The
change [in the new allocation key for eastern bluefin tuna] was made
without any preceding agreement . . . or any discussion in Panel . . .
or in the Plenary regarding the criteria for such changes."[48] I learned
that Algeria was the only other ICCAT member state to formally object
to ICCAT Recommendation 10–04, which was not surprising given the
drastic reduction in its share of "stock."

Two years later, at the commission meeting in Agadir, Morocco,
the Algerian delegate candidly said to me during a coffee break that
the quota reduction was an "injustice," a matter of "national pride," a
"wrong" against his country. He felt Algeria "a victim" of the institution, a country that "voluntarily" submitted to ICCAT rules but that
was nonetheless powerless to make a change without the consent of
other member states. His hands tied by the commission, the diplomat
revealed that ministers of state and members of the press in Algiers
were eagerly awaiting news. Would the commission agree to return Algeria's share of quota now that member states were again renegotiating
the allocation in Agadir? Without a formal dispute mechanism at the
time, ICCAT left him and the fishers he represented in the lurch. To
take the dispute to the World Trade Organization was a course Algeria
was unwilling to pursue.

In the end, nothing changed. The commission did not agree to
reinstate Algeria's share of export quota in bluefin tuna at the Agadir
meeting. As one delegate of state said matter-of-factly to me, "Algeria
got hosed."

"Game" Over: The Empire of a Global Regime of Value

An ICCAT meeting was a privileged site for some actors to perform, through the play element embodied in the ICCAT "game," the calculative dimension necessary to secure their investment in "red gold" by adjusting to a legal situation limiting supply and market access. This was an arena where member states—especially if rogue and rich—carved out authority over who controlled oceanic space, and the fish left in it, so that they might render the export markets of ICCAT signatories legal. The Atlantic bluefin tuna was the material and symbolic force that mediated this activity.

Although intergovernmental organizations such as ICCAT may have assisted in reducing raw, physical violence on the high seas, the technocratic posture of supranational rule making does not mean that violence has been eliminated entirely. The rapid, legalized slaughter of fish post–World War II attests to this. The "injustice" experienced by Algeria demonstrated that coercion only became noticeable when ICCAT was confronted with a breach or an unfair, shocking instance of exploitation.[49] The Algerian case was more than another instantiation of the Global North strong-arming the Global South. Nation-states—many of them former colonies and protectorates—must accept what they otherwise would not because they voluntarily participated in supranational regulatory agreements to access global markets. Participation in ICCAT was a requirement for countries to survive in a geoeconomic age.

Much of what passed for economic activity at ICCAT was not dedicated to the creation and promotion of the so-called free market. The presumption that industry was unhinged from government was a fantasy. Delegates did not pretend that state and industry were separate, hermetically sealed spheres, nor did they enter the playing field to compete as equals. Instead, the delegates working for well-financed member states spent much of their time enacting or escaping potentially coercive situations as law was in process. Contests were for prestige as much as they were for commodity empires. The caper orchestrated by Libya before the fall of Gaddafi showed the great disparity among delegations in the freedom to maneuver.

Instead of reducing the ICCAT regime to brute economics or to politics writ large, this chapter has shown that the stakes reside *equally* in the realms of the social and the cultural. The status of ICCAT as a supranational regulatory regime, the status of a delegation, and the status of an individual delegate mattered as each tried to secure its position in relationship to others. To perform institutional authority, to aspire for commodity empires, to assert resource nationalism, to administer a functioning state, to advance one's own career, to exhibit deft and nimble negotiating skills were all brought into play here. The range of stakes explains why delegates returned each year and recommitted themselves to the "game" for its social reproduction. To play a role in the drama, to reenact the ordered ritual, to be a spectator in the match helped to bring delegates back to repeat this tournament of value every year—without prior knowledge of the exact outcome. This applies both to previous champions and to chronic underdogs.

Powerful member states built or retained their commodity empires through ICCAT's bureaucratic apparatus. Rogue petrostates extended their market share in fish through clever plays, to be sure. But what I think I was also witnessing was the empire of a supranational bureaucracy, or what José Alvarez calls "the empire of law."[50] As ICCAT member states claimed their share of the pie, the bureaucracy defined the architecture for the field of play. In some respects, it was from the very structure of the "game" that the asymmetry was begotten: the rules of procedure, the mass of bureaucratic paperwork, the relatively small number of official spoken and written languages, the size and composition of delegations, the learned diction of the diplomats, the mastery of fisheries science, even where one sat in the room when meetings were in session. These elements of the "game" sorted delegations into hierarchical rank and favored the well financed. It should come as no surprise that the process through which member states made decisions—consensus—was shot through with, and entangled in, webs of power.

Must consensus imply unanimity? How can international organizations better process dissent? It would be missing the point entirely, it seems to me, to call for a return to the voting system, as if this alone would solve the problem of inequity.[51] At a time when commodity empires under finance capitalism have captured apparatuses such as ICCAT, the power of law in process implies that changing the paint of a room will not materially alter the structure of the building.

Ocean governance as currently constituted is not the silver bullet that will quickly solve the overfishing problem, let alone redistribute wealth, mitigate economic injustices, and resurrect a once giant tuna. The ICCAT apparatus has become a structural feature of commodity empires under finance capitalism. Despite the extraordinary demise of big fish under ICCAT's tenure, member states have been doing the job ICCAT was founded to perform a half century ago: to protect the export markets of signatories under the false presumption that the inventory in high-seas fish was inexhaustible and could be engineered for economic growth.

Where do we go from here?

Conclusion

〜〜〜

All Hands *on* Deck

CONFRONTING *the* SIXTH EXTINCTION

On a cool damp night in November 2013 my alarm rang at half past two. It was time to visit Tsukiji, the famed fish market-place in the land of the rising sun.[1] Tokyo travel guides advised tourists to arrive early to secure one of 120 spots for the tour of the tuna auction in Japan. Here in the refrigerated cold, intermediaries supplying the global trade in sea creatures met buyers in the morning dark. Although the International Commission for the Conservation of Atlantic Tunas (ICCAT) maintains its secretariat in Madrid, the hub of the sushi economy was Tokyo. The scene of dead bluefin tuna lying side by side in ordered rows on slabs of wet concrete like chalky tombstones in an indoor cemetery formed an image seared in memory. The deceased specimens were grisly reminders of the material exercise of power over nonhuman animal beings. Gutted, the bluefin as "red gold" embodied the violence of commodity empires in the global ascendance of monetized value.[2]

But this image, I believe, needs to be juxtaposed with another: the incredible Atlantic bluefin tuna whose enormous heart propels her to jet between continents, playing with mates across undulating waves below the ocean surface, amid the chatter of various sea creatures, affirming the wonder of life in unmatched migrations the intrinsic value of which is incalculable. This image defies the organized effort to scientifically know and rationally plan her extermination, systematically

Figure 14. Innards gutted, dead bluefin tuna from Okinawa, Japan, awaited purchase at auction in the frigid holds of Tsukiji marketplace in November 2013.

executed since the 1950s through the statistically derived, juridical formula of "maximum sustainable yield," steeling commodity empires.

Tsukiji is regularly ranked top on lists of Tokyo destinations in travel guides. After visiting the tuna auction, the fish stalls, the retail shops of Tsukiji, I took a ride on the Tokyo rail. To my surprise, it was a site less trafficked by international tourists that struck me most that day.

Tokyo Sea Life Park was the first aquarium in the world to display tunas publicly. It opened in 1989.[3] The bluefin brings considerable prestige to this venue. In fact, the entire facility was designed, says its Web site, to exhibit the fish in "a huge donut-shaped tank (2,200 t[ons]), where bluefin tunas swim around freely."[4] Far from the congestion of Tsukiji, near Tokyo Disneyland, this seaside aquarium, on the day I was there, attracted mainly Japanese nationals, families, and school groups. Circling in a spare holding tank, the bluefin did not seem to me to swim freely.

I wanted to meet the protagonist of this book alive, no matter how staged the circumstances of my encounter. I wanted to behold the moving, breathing Atlantic bluefin tuna, even if in fact this was her Pacific cousin. Only a few people have ever seen her underwater firsthand, in the open ocean, schooling, shy, I imagined with her occasional traveling partners, the yellowfin and the dolphin, coordinating hunts and diving three-quarters of a mile down to depths where ocean temperatures approach a few degrees Celsius. The closest I had come to her before had always been near her impending death: aboard the recreational fishing boats I knew from childhood and while on watch observing pens at a "tuna ranch" I visited in the Mediterranean during this research project.

I missed the tuna feeding session. Contrary to the popular imagination, what happens at aquariums and their land-based counterpart—the zoo—is not about the "animal." It is about people. Feeding rituals hardly resemble natural behavior and instead cater to spectators who watch as if entertained by "the animals' difficulty in eating, titillated by the visceral incontinence of their gross appetites."[5] Born of the same culture as ICCAT, Tokyo Sea Life Park was entwined in histories of commodity empires wherein the attempt to master and control sea creatures "risks foundering amid our imperious ecological ethos."[6] That the movement for marine conservation gathered speed at the same time as the overexploitation of fish after World War II is not coincidental. They are two sides of the same coin now called the Great Acceleration.

Jet-lagged and disoriented, restless and eager, I rounded a corner in a dark chamber of the aquarium. When the bluefin came into sight swimming through a "violet-black" ecology[7] in this cramped artificial sea, I began to cry, overwhelmed, like a pallbearer in mourning at a funeral. I had not expected this response.

Tears fell down my cheeks as I wept silently on a hard bench positioned for onlookers to rest while staring at the bluefin's customized pool. I peered motionless through the glass separating me from her: so close yet so distant the bluefin was from me. She could not possibly know I had spent the previous four years riding her tail across the ocean, through millennia, running interference on her behalf, or so I imagined. I wanted to thank her for allowing me this journey, but my words would never reach her.

She was there in the dozens. She was smaller than I wanted her to be, even though I knew the giants swimming in packs were no more. Changes in the sea as a result of commercial activity had been

Figure 15. Inside Tokyo Sea Life Park, dozens of bluefin tuna circled indefinitely in November 2013, a year shy of a mass die-off when only one remained in the tank.

happening since at least the end of the seventeenth century. The warning signs have been evident in each new generation since. Baselines have shifted, bringing with them collective amnesias.[8]

I watched her streamlined body glint as light reflected off her tiny scales. Every so often a bluefin accelerated in a burst of speed. All flashed like polished chrome in the sun.

"Silver and gold. They rule the sea," so read my field notes from that day.

I wondered: Was she bored swimming in that donut, round and round, unmoored from life pelagic, only a short distance from the gateway to the vast ocean expanse via Tokyo Bay? Could she adequately exercise those formidable muscles when swimming, enclosed? Did she have enough to eat? Did the artificial light bother her? Could she see and hear us? Was it too cold, or too warm for her in there? Did she carry cancer in her body? I remembered a marine scientist who told me about tumors caked in the muscles of the Mediterranean swordfish people eat.

Or was I supposed to congratulate her keepers? At least here she was not sold to the highest bidder at auction, chopped up and distributed to all corners of the globe, served as fashionable meat to stoke the vanity of the leisure class. At least here she was not choking on the oil and dispersants that seeped into the sea as a result of energy extraction, as was the case with her kin in the Gulf of Mexico who dodged the blown Deepwater Horizon oil rig. At least here she was not radioactive from the nuclear waste that surely passed through her gills after Fukushima. At least here, like a museum piece, she was safe from an ocean becoming warmer, more acidic, more polluted, less oxygenated, less biodiverse. Or was she?

Should I feel as trapped as I imagined her to be in this dizzying world of marine conservation? Which way was out, for both of us?

As I departed the aquarium, outside the gift shop where visitors could purchase stuffed velveteen dolphins and tiny plastic sharks glued to magnetic strips, I passed in the lobby a large plastic replica of a single bluefin tuna. Her head faced the ceiling. A metric ruler showed in increments how big she gets if left in the sea to grow. Dwarfed by this giant, children flanked both sides of this fakery for a photo opportunity. I looked from a distance at three girls beside this synthetic beast. Perhaps they were sisters. The eldest I guessed was ten. Two raised their fingers in the universal sign for peace.

More than a year later, in March 2015, my sister sent me a news article in the *Washington Post*. The headline: "The Mysterious Deaths of All but One Bluefin Tuna in Tokyo Sea Life Park." All but one of these captive beings had died over the course of four months. One bluefin tuna was left to swim alone in a tank emptied of all other inhabitants. Some of her companions suffered broken backbones once they crashed into the acrylic wall, which, said an aquarium representative in the article, "was unfortunate but not very unusual for tuna kept in a tank." Sixty-eight bluefin tuna had perished. It was the aquarium's first mass dying. The article asked: did a virus cause their death, or were these creatures too sensitive to light and noise? The author ends the piece: "News reports earlier this year said the remaining fish seemed unusually nervous and wouldn't eat."[9] Did the bluefin experience grief or commit mass suicide as other creatures do?[10] Thoughts of one bluefin tuna isolated from her kin in that monstrous tank gutted me.

"Why are these animals less than I believed?" John Berger once asked of the zoo experience.[11] He meant not that nonhuman animals were inferior or subservient to people but that the experience of seeing them caged tends to disappoint and asphyxiate the spectator. Creatures cannot possibly live up to an onlooker's expectations of the wild when they are condemned to a life of incarceration. Like other confined sea creatures, the bluefin tunas were supplied food at regular intervals and were housed in a manufactured saltwater tank filtered by electric power. I think the disquiet she provoked in me that day was because I saw a fellow being on view, rendered before my eyes as, in Berger's words, "*absolutely marginal*; and all the concentration [I could] muster [would] never be enough to centralize it."[12] That her exploitation was social—and not just economic—was the scandal here.

So detached, estranged, and out of touch have most people become from their place on Earth that those closest to nonhuman animals have fled to the margins.[13] The meaningful encounter between people and nonhuman animals as cocreators in this life has played an absolutely crucial role in experience until only a few centuries ago. It has now for the most part been extinguished by the modern condition.[14] Species did not "meet," "touch" or engage at the aquarium—or at ICCAT.[15] Like a zombie, the bluefin was evacuated of life but was alive with power as a commodity for the profit and the consumption of global elites. "Red gold" was as much a condition as it was a biological outcome of new techniques of sea power.[16] This was not inevitable.

The innovation these institutions claimed was compensatory—making up for what has vanished. But in reality, these institutions belonged to the same remorseless movement that created them and that now gives them purpose. Like the aquarium, under the veneer of "civilization," ICCAT has become a living monument to sea creatures disappearing from the everyday life of the vast majority of people.[17] But invoking nostalgia is not my intention. What has disappeared is not only the last of the truly giant bluefin tuna, or how many of them there once were in a habitat that allowed her kin to flourish. Lost is a mode of relating to beings—to all beings—in the modern social and ecological experiment called "progress" leveraged for commodity empires and administered by states under extractive capitalism. Entire institutions have been formed by—and remain faithful to—the myth that to be "civilized" is to be separate from Nature. The sixth mass extinction is the planetary consequence.

Treated as a machine running on instinct, devoid of sociality and consciousness, the bluefin has been conceived by elite "improvers" in high modernity as an isolated unit and indifferent commodity passively awaiting consumption. She has been stripped of history, majesty, and vigor to become a technoscientific object for bureaucratic scrutiny. She is not a "very elder god," described in the Prologue of this book, but a delicacy manufactured for the wealthy and an entertainment sold to children. This nonhuman animal being has become an icon of fabricated demand utterly dependent on the unrestrained wills and desires of her keepers. The technocrats in ocean governance need the bluefin to survive no matter how small she has become, not because they are stewards of the planet but because they are wardens of economies attuned to commodity empires for the accumulation of wealth. The statistical models determining how many fish to keep alive for export rationalize in policy recommendations this "development."

Anna Tsing calls this condition "Earth stalked by Man." Earth. Stalked. By. Man. Relations created by ecological simplifications such as the plantation have transformed living beings into slaves, into "resources," into biological assets, removing them from the complexity of their lifeworlds for their replication. But to make such simplifications possible aboard the factory vessel or at the "tuna ranch" required the work of culture. "Let's call this work 'alienation' whether it involves humans or nonhumans," Tsing writes.[18] Alienation in fisheries management has implied since its formation a regulatory culture valuing living beings above all else as replicable assets inventoried as "stock."[19] This worldview, this disposition, this orientation to other living beings helps to "produce that model of the future we call progress."[20] Build another holding pen. Count another beast. Tame the remaining schools of predators. Inventory the lot. Treat competition and greed as natural to species being. Mark countries as "civilized" under international law. Believe that technology will save us from the sixth mass extinction.

"Ideas of nature," Raymond Williams writes, ". . . are the projected ideas of men."[21] To render a nonhuman animal being first and foremost a commodity—to evacuate the soul, the sentience, the mystery of a living being—is to project onto the perceived darkness of the deep sea "man's" own ideas of a singular, abstract, absolute Nature. Not all cultures across all time and place have related to nonhuman natures in this way. In such zones as ICCAT, many delegates have become detached from the natural environment, as if Nature is another abstraction already endowed with consistent and reconcilable laws awaiting scientific discovery. The technocratic establishment favors the treatment of the environment in this way. But these ideas—emblematic of predatory regimes of value—in their narrowness write out of existence much too much history and obscure the assumptions that underpin the dominant interpretations of experience in society.[22]

An ocean giant has been put to death by a thousand cuts in the name of marine conservation. Interventions based on appeals to economic and scientific rationality have become so overexercised in the pursuit of sea power that nonhuman natures, in spasm, are now as much a source of volatility as is marine policy.[23] The finance capitalists, under the protection of an administrative state scaled supranationally, continue to accumulate wealth no matter how few wild fish are left. Like escape artists, financiers assume that investment in aquaculture will release us from the harms of overfishing. Fish farming has grown exponentially over the last few decades, so much so that it now accounts for nearly half the world's fish supply.[24] Why confront head-on a predatory regime of value, if an ill-informed public continues to believe in the cornucopia myth, as if there is always another fish in the sea to buy?

To solve a problem of their own making, technocrats performed rituals of order at ICCAT meetings. Their precise resolutions allocating quotas for bluefin tuna said to cluster in the eastern Atlantic extended to some forty pages. Their "Kobe Matrix" captured fish "stock abundance" in probability estimates.[25] Their rules of procedure favored and assumed consensus.

Yet, arranged in the world of ocean governance are not only sea and fish *but* people.[26] The technocrats at the controls are not seriously considering what their reshaping does to sea and fish *and* people. The task must not be to create another abstraction or promote another techno-fix. Needed are honest, on-the-ground accounts of the efforts, seemingly benign, to invent, produce, and fortify the divisions between living beings based in a predatory logic expressive of how hollow and how broken relationships have become.[27] Without addressing this problem, I am afraid, the solutions proposed will continue to be short-sighted and short-lived.

Back at the aquarium, the bluefin "look[ed] sideways" through the plate glass, "blindly beyond," "scan[ning] mechanically," "immunized to encounter."[28] Unable to engage with the gaze of the being on the other side of the glass, I once found myself similarly alone.

Who was really caged at the aquarium—or imprisoned within ICCAT?

I do not expect ICCAT, if it remains attuned to an exterminatory regime of value, to "save" the bluefin and restore her giant kin in the midst of the sixth mass extinction. Under the prevailing conditions of valuation, ICCAT member states are not equipped to radically alter international environmental governance on their own. The ICCAT regime is so entangled in global webs of administrative power and wealth accumulation that it has become a structural feature of commodity empires.

It is abundantly clear that overfishing is the result not only of the (in)action of one lone regional fisheries management organization (such as ICCAT). Rather, overfishing is the result of an entire culture that takes as one of its expressions the current system of ocean governance wherein the commodification of life through law and fisheries science is the modus operandi. ICCAT is representative of a host of specific habits of being flowing from a complex system of alienation spread throughout the social structure. That system affirms through the sociality it promotes a regulatory culture that values life in one way—and not in others.

This book is not meant to be an indictment of the individuals who pass through or work for the ICCAT apparatus. The lesson should never be contempt for the people working within the only organizational space available to protect high-seas fish in the Atlantic basin. The fish slaughter legalized through the ICCAT regime requires much more than assigning blame to the ICCAT insiders who drank the Kool-Aid of "maximum sustainable yield" and "electronic bluefin catch documents" (e-BCDs) in fisheries management. Many of the delegates I met were well intentioned and hardworking. They were aware, even officially, of ICCAT's shortcomings.[29] Some shared with me their disappointments, their frustrations, their regrets. In ICCAT they had found their career, their reputation, their whole life's work. Some did their best to conserve sea creatures, as they understood it, under conditions not entirely of their making. Some believed they were doing right by promoting economic prosperity for their country. All were entangled in the great forces of international law, finance capital, nation, state, fisheries science, technology, and civil society, which had been brought to bear on a highly valuable, mobile creature extracted from waters many fishers once trusted to be inexhaustible in their riches. Many ICCAT delegates were at the mercy of unseen forces that were completely unknown to them. The bureaucratic machinery demanded that they play the "game" of economic growth by determining who knew what when. Such are the ways political operators leverage social and economic capital across sectors in the technocratic establishment today.[30]

On occasion I see interlocutors I have met through the ICCAT apparatus pass through New York City for meetings at the United Nations. One wanted to know my conclusions, my policy recommendations, my insights into how I would respond should the press call looking for a sound bite. How would I answer these questions in twenty seconds or less: Has the Atlantic bluefin tuna "stock" recovered? Did the environmental campaigns have an effect? Is there reason to celebrate a second (or third) revival of the species?

I worry these questions maintain at their core the fiction that the technocratic establishment as currently constituted is equipped to assume control over the planet's destiny in a way that would adequately address the coming catastrophes of the sixth mass extinction. The hubris inherent in the "civilizing" project to master and control Nature has allowed global insecurity to rage unchecked. It takes an extraordinary amount of work and coordination on the part of ICCAT delegates— indeed, it takes an entire apparatus—to give the appearance that expertise evacuated of soul can overcome the fluid things of Nature.[31]

Following the lead of Audre Lorde, we must ask ourselves if the master's tools will dismantle the master's house.[32] Can their representatives find the portal to exit this predatory regime of value on their own? If they can, how long will the renovation take? Iver Neumann poignantly shows through an ethnography of the Norwegian foreign ministry: "The focus of diplomacy is maintenance, not change."[33] "Make haste slowly" goes the idiom.[34] Environmental diplomacy cannot be expected to realize the radical, rapid change required to ensure a biodiverse ocean. Yet we need multilateralism now more than ever. Cooperation across borders is basic to charting a course for shared futures. What kind of global architecture would ensure that decision makers care for long-term planetary futures as much as they care for short-term, national, economic ones?

Some people talk of accountability deficits and a lack of transparency in a global administration without a constitutional basis for checks and balances.[35] In this view, democracy and civic engagement are needed to combat dense bureaucratic labyrinths and the seizure of administrative law happening on both domestic and supranational scales. Others say that to "save" the ocean, more science and more technology are needed. Techno-fixes will save us from ourselves!

Critical histories of fisheries science for fisheries management are urgently needed, as chapter 4 made clear. Science cannot be trusted when monied interests have captured it. The aim of critical inquiry must not be only, in the words of Isabelle Stengers, "to uncover . . . the conquering machine [the scientists] conceal." This is where deconstruction alone would take us, leaving "a more desolate, empty world" in its wake. In her "manifesto" for "slow science" the scientists would become freed "from the powers that have subjugated them," leaving them to pursue the creation of "relevant modes of togetherness, between practices, both scientific and non-scientific; finding relevant ways of thinking together."[36] Will the fisheries science applied in fisheries management be able to release itself from the industries it is supposed to help regulate, thereby enabling this mode of engagement?

Scientific understanding about the lifeworlds of sea creatures must be embraced and promoted—if hinged to respecting their being.[37] Perhaps one day marine policy will accommodate the evidence that fish feel pain;[38] that aquatic beings show their individual personalities when responding to stress;[39] that certain sea turtles, sharks, and cuttlefish communicate through bioluminescence;[40] that the cleaner wrasse recognizes herself in the mirror, considered "the gold standard for determining self-awareness;"[41] that fish sing at dawn and dusk like birds when carrying on their social lives,[42] that the intensity of their chatter may help us know how many of them there are.[43] The experts I met while traveling through the ICCAT regime were not open to this way of scientific thinking and how it could transform them and the marine policies they developed. *More scientific facts will not necessarily alter the values that underpin how those facts are arranged.*

The problem of overfishing in times of extinction is so massive as to call into question what law and fisheries science can do. This does not mean they can do nothing. We face, in the words of Giorgio Agamben, "the powerlessness of men, who continue to cry 'May that never happen again!' when it is clear that 'that' is, by now, everywhere."[44] "That" I take to be the traces modernizers have left in the ocean. They are everywhere, even in the deepest crevice found seven miles below the surface of the Pacific Ocean in the Mariana Trench where synthetic fibers contaminate the stomachs of tiny sea creatures luminescing in the dark.[45] Regulations and scientific studies, as important as they are, cannot, under present conditions, do all the work.

I worry when the establishment places all its bets on a techno-fix, on more "dreamworlds of progress," on well-worn masculinist fantasies of the sword.[46] In this view, proponents of the "blue economy" claim they can monetize marine "resources" still further by highlighting the asset values of "natural capital" so as to show to humanity the dividends in store. Consider that international legal instruments propose to treat fish not only as commodities but also as "marine genetic resources" ripe for profiteering. In derivative finance, investment in fish as genetic sequences becomes an opportunity for global elites to hedge, measure, and make legible risk in the next frontier of extractive capitalism.[47]

Pundits laud the latest technologies associated with the fish trade. The World Economic Forum has weighed in and declared "tuna traceability" as the way to curb pirate fishing.[48] I wonder: How will poor coastal states participate in this techno-fix? Must they develop ports and reliable electric-powered infrastructures required to toy with tech? Who will pay for—and profit from—their implementation? Will high-tech tuna benefit only the wealthy, leaving the poor still farther behind? Will the captains of industry swindle the consumer-citizen yet again by passing on the cost, because fish will become more expensive from these investments? What other "golds" will emerge—red and white—as the criteria for trading scarce fish becomes more complicated?[49] Must we live in a society structured so that scarce fish remain worth stealing? Must we turn a blind eye to forced labor in the fishing industry, which keeps the price of fish artificially low for consumers in rich countries?[50]

Some say that if change cannot happen from within, then outside pressure from the public is needed. The most forceful efforts to restrain regulators from overexploiting fish have come from mainstream environmentalists. Chapter 3 showed that the narratives wed to the savior plot compelled ICCAT to pull back on "harvest" in the short term. The attention the news media paid to Atlantic bluefin tuna once pressured ICCAT member states to more tightly regulate the extraction of this one fish. Yet, even if the bluefin tuna "stock" has "recovered," even if the giants returned, can we assume other fish have also been "saved"? When rhetoric treats the bluefin as a canary in a coal mine, as the part representing the whole of the overfishing problem, it becomes hard to specify what has happened to other creatures in an overstressed ocean, especially the ones on which poor fishers and their families depend.

For whom, we must ask, has the bluefin been "saved"? *From what* has she been "saved"? The bluefin has not been "saved" for her intrinsic value or for the niche she fills within indispensable ocean ecologies. She has not been "saved" for poor artisanal fishers or even for wealthy recreational ones. She has been "saved" to preserve the export markets of well-financed ICCAT signatories and the power on the world stage derived from them. *Sea creatures need to be respected and their mysteries widely shared for them ever to be "saved."*

This book should absolutely not be read as a call to gut and abandon institutional efforts to stem environmental catastrophe, including those emanating from the diplomatic realm.[51] Institutions and the regulations they produce are not inherently bad. I hope this book instead prompts the reader to ask of ICCAT and the network of institutions in environmental governance: To what purpose are they directed? To what values are they oriented? To what culture and history are they indebted? In which claims on wealth are they entangled? How can the logic of profiteering be dislodged from multilateralism? How can multilateralism promote ways for these institutions to extricate themselves from predatory regimes of value? What new social contract can these institutions help bring into being?

My argument to confront continuity in predation over time accentuates the enormity of the task ahead across all sectors of society. To acknowledge a crisis as truly systemic means no one is off the hook. It requires at minimum that we zoom out, no longer compartmentalize, orient ourselves to the long term, remember our history, listen to a variety of voices, and compel change big and small from all. To spare oneself the effort of looking at the whole complex of relationships will only further the alienation. Evasion opens the way to supporting processes of totalitarianism.[52]

The lesson is not to ditch and recoil from the whole of the Enlightenment tradition because ICCAT is not working for all beings on this planet. As stewards of the ocean in a time of extinction, people need reason now more than ever. But reason alone is not enough. ICCAT would be a very different institution if the capabilities of all beings, including those of the sea, surprised and impressed us. More often than not global elites "forget how they themselves are rendered capable by and with" those creatures called "animal."[53]

Many people within and beyond the world of ocean governance find it hard to reckon with—and intrinsically value—the nonhuman animal being-in-the-world. It is too disruptive under the present conditions of political economy to administer commercial fish knowing, for example, that Atlantic bluefin tuna defies common categories. She is warm-blooded. It is too unsettling for decision makers to act on the knowledge that in her thunderous speed, like an underwater ballistic missile, the bluefin can cross an entire ocean and explore aquatic places in the Sargasso Sea the human eye has never seen. It is too destabilizing to the status quo to reckon with a lobster being, knowing that she can detect small changes in ocean temperature through thousands of tiny hairs and that her nervous system still functions for an hour after a cruel death once scalded in pot. It is too troubling for modernizers to reckon with a three-hearted octopus being. She uses tools without a bone in her body and is endowed with memory, recognizing caregivers through lifted eyelids while caged in tank. She is so bored that aquarists must invent ways to keep her busy as she changes color, pattern, and texture in seven-tenths of a second.[54] Sea creatures had to be devalued and reduced to their bare life functions—eating, defecating, hunting, making little babies, if they can still find a mate—for them to be valued first and foremost as biological assets, as "stocks."[55]

Technocrats cannot regulate a way out of a mass extinction they have helped to create. Without addressing this sobering fact, regimes such as ICCAT will continue to maintain *the fiction that the technoscientific ordering and capitalist values they hold dear are separate from the physical collapse of a deteriorating ocean.* Alternative modes of engagement—those beyond salvage conservation and "maximum sustainable yield"—that lead to a revaluation of life are urgently needed. In the words of Williams: "I think nothing much can be done, nothing much can even be said, until we are able to see the causes of this alienation of nature, this separation of nature from human activity."[56]

Some readers will reject this stance. Some will be unable to occupy the place from which to know the world and themselves differently. This is "idealism," I have heard some of them say. What we need is "pragmatism" for "practitioners" to do the real work of marine policy making. These are the experts who will find it difficult to see that they, in fact, practice their own idealism by refusing to confront the ways in which they are bound to a predatory regime of value. They and their funders have the most to lose from dislodging this value system. They might ask, "Where are the policy recommendations in all this? What concrete 'solution' does your book offer?"

To me, more probing questions would ask: Must we live blind to the histories described in this book? For those who have been touched by them, how might we release ourselves from, and become more accountable to, this heritage?[57] What would policy worlds unmoored from the commitment to "progress" look like? Might more touch, more respect, more courtesy, more regard, more curiosity, more humility, more engagement, more shared attachments enable movement along a path where creatures meet, as Donna Haraway has encouraged? Might an approach as modest as no longer regarding a sea creature as an "it" be a very small start, as this book has tried to model?[58] I am not prepared to say *what* is to be done as much as I am prepared to say *how* to get there.

The world is more than a problem to be solved. It is also a mystery to be contemplated.[59] There is no wonder drug or magic bullet equipped to forestall the ruinous change this planet has rapidly endured since World War II. The problem has expanded beyond the scope of what one lone solution can accommodate. Ocean governance was born of the violent rifts of a savage political economy spreading across the globe to produce marginalized, disposable beings of all kinds, human and nonhuman. In the words of Thom Van Dooren, "[O]ur inability to really *get*—to comprehend at any meaningful level—the multiple connections and dependencies between ourselves and these disappearing others [is] a failure to appreciate all the ways in which we are at stake in one another, all the ways in which we share a world. This failure is, at least in part, rooted in human exceptionalism."[60] Human exceptionalism has been stitched into the very fabric of the institutions entrusted to care for life in the ocean. A transformative vision of a peaceful, fair, equitable, related world in which elites are no longer the center of the universe is required to go further and more boldly into the Anthropocene. The time is now to lay the foundation and establish the conditions for this just transition.

In 2012, my father sold his seventeen-foot *Aquasport* to a family friend for a few hundred dollars. He once woke up at dark to catch first light at sea. That's when fish rise with the morning sun and bite. Striped bass was his game. He and a neighbor or two would ride back on his center console and, after docking the boat, announce their catch. As time passed over the years, my mom and I heard more frequently the report: "Chasing phantoms." By Shinnecock Inlet on eastern Long Island, anglers now chase phantoms where once they chased fish. ICCAT did not create this problem, but the world ICCAT exemplifies did.

Phantomlike, barely surfacing, the tremendous Atlantic bluefin tuna haunts me as she does this book. In a predatory regime of value, the once giant bluefin tuna has become a zombie: present but absent, smaller, exhausted, unable to contribute to vibrant ocean habitats as her kin once did. "The winds of the Anthropocene carry ghosts—the vestiges and signs of past ways of life still charged in the present."[61] Ghosts in the machine, spirits leaving traces, disembodied, soulless. They orbit consciousness in an "imperialist nostalgia" that mourns for what one has destroyed.[62] To confront this social fact directly is required to neutralize its power. To address the root causes of harm is a precondition for a way forward.

The apparition of the giant bluefin tuna reminds us that, like all other prior systems, still, this one too can be undone. Routes of solidarity are emerging from which to imagine, respect, engage with, and care for others in this scarred world.[63] Despite the pollution, the acidification, the dead zones, the die-offs, the underwater heat waves, I hope the reader is not left paralyzed or alone. The status quo yearns for that response. Collective action is needed. The worlds brought into being by such institutions as ICCAT are not timeless and universal. They will loosen their hold, become unstable, and be reconfigured. Nothing, history reminds us, is forever. New meaning will emerge from the ruin. But which meaning will seek to impress itself and carry the day? Who among us will commit to the enormity of the situation, by resisting this predatory regime of value and shaping a future not yet decided? And how can wisdom based on the mutual dependency of all life help reorient the institutions we desperately need to care for our common home?

All hands on deck.

Acknowledgments

~~~

November 26, 2009. Thanksgiving Day. Not a cloud in the sky. Light wind. Glorious. I had a pocket of time midmorning before I turned to helping my mom prep for the holiday dinner. Late fall tends to bring sea breezes strong enough to stir whitecaps on the bay, but not this Thursday. I suited up and went for an unexpected kayak run on Shinnecock Bay, where the south shore of eastern Long Island meets the Atlantic Ocean by barrier beach in New York State. The still water, the crisp air, the brilliant sun, the crystal blue sky were gifts to this former dissertator unable to settle on a research topic until this day.

The summer boat traffic had subsided by now. No one else was on the water. I paddled past the osprey nest vacant this time of year. It towered over a nature preserve once the home of the Shinnecock, "the people of the stony shore." The thousand-acre reservation for the Shinnecock Indian Nation lies on the opposite side of the bay, to the east, cast away, down-market in relationship to the stretch of lavish seaside estates celebrated in American culture.

Not far from shore I took a break from paddling, and drifted. I did not know then that this moment would send me on an errand lasting more than a decade. Although I was aware of how deteriorated the sea had become, I felt it that morning. Intensely.

Where did all the eelgrass I knew from childhood go? There had once been a patch so thick by the house that during low tide the tips of the grass would peek above the waterline. They would tickle my leg and send shivers down my spine if I swam too far from shore. That eelgrass is now gone.

I looked down. The bay was clear enough to see the sandy bottom this time of year. Nitrogen runoff from the septic tanks and the fertilizers greening the lawns had declined because the summer crowds had left. I sat, motionless. There were no immediate signs of life. No razor clam shells. No horseshoe crabs. No schooling baitfish.

Less than two weeks later, in December 2009, the United Nations Climate Change Conference was under way in Copenhagen, Denmark. Dangerous carbon emissions, damaged rain forests, and diplomatic stalemates made global headlines. Where was the ocean in all of this? I had heard somewhere in the news of the plight of the bluefin tuna and was vaguely aware of the International Commission for the Conservation of Atlantic Tunas (ICCAT) from the articles appearing in the sport-fishing magazines my brothers shared with me.

My grandparents first came to Shinnecock Bay in 1956, a time consequential to this book. They accumulated enough savings to buy for $2,500 a small, sandy lot they saw in an advertisement in a local newspaper. The Hamptons was not yet known for its bling. Little by little, Nanny and Pa built with their own hands a modest summer "cottage," as Nanny would say, helped by family and friends. Not until my sister's arrival in the late 1960s did they install a landline and heat. The television came in the early 1980s. Instead of idly watching the "idiot box," Pa said, we went fishing for fluke and flounder. He joked about how he thought I was his good-luck charm because the fish would bite when I was with him. It was hard to explain the behavior of wild fish.

Of course, fishing with Pa was not my only memory. Local anglers were characters: some loud and salty, others reserved and quick-witted. A few had a cocktail and cigarette in hand by five o'clock in the morning, which enlivened the fish tales they told in the bait-and-tackle shop my father regularly visited. The most colorful was Frank Mundus, so legendary in Montauk that he was said to have inspired the character Quint in the Peter Benchley novel and Steven Spielberg breakout film *Jaws*.

Not far from these channels sushi went global during my childhood years. Over time, the scene exploded with McMansions, Jet Skis, helicopter racket from the Wall Street executives dodging overhead all that car traffic, and tiresome displays of wealth so great that a compound larger than the White House hides behind the hedges owned by a junk-bond billionaire. In all the stories about the lifestyles of the rich and famous, of the gala dinners and movie premieres, of the scandals involving the who-done-it murder at an oceanfront castle, and the publicist who in a rage ran over clubbers with her luxury SUV, something was happening in the waters all were drawn to: a sea change so momentous that only recently are people coming to terms with its consequences and, I hope, their own complicity in it.

As the writing of this book was completed, *Newsday* published this headline, days after the Labor Day holiday weekend on September 5, 2019: "Thousands of Fish Found Dead in Shinnecock Bay." My dad called them "peanut bunker." It was the first mass die-off of juvenile menhaden western Shinnecock Bay has ever seen. It happened where the resident osprey now collects shredded plastic bags to help fortify her nest, where the disappearing eel in her grass warned us that ruin would come years ago should ignorance, indifference, and denial continue. Runaway pollutants leached into the water after a soaking rain. The fish did not have enough oxygen to breathe. Seagulls, my dad said, were "having a feast." The reporter quoted Southampton Town officials who downplayed the severity of the event, as if to neutralize public alarm. Who would take responsibility for a creeping dead zone caused by rampant real-estate development and poor waterfront planning, compounded by climate catastrophe and rising seas? There was too much money to be pocketed somewhere along the line.

This book acknowledges the impending loss of a social ecology that has nourished my family and friends for decades. My grief, my indignation, my anxiety, my torment are not new or exceptional to the human experience. I remember the Shinnecock who have summoned the strength, wisdom, and tenacity to endure over the centuries.

While this book appears to have only one author, in fact there are many people who have written it and to whom I am warmly indebted. First and foremost, I thank the ICCAT delegates in the field who generously gave me their time when sharing their experiences. They answered my questions with patience and goodwill. Without them, I could not have possibly found my way through the fog of ocean governance. I respect their anonymity and have tried to represent and synthesize their views as best I could. This project qualified as an exempt study in review by the University Committee on Activities Involving Human Subjects (NYU HS# 10–8162).

This book takes as its compass the work of some of the greatest minds in the academy. When I was a doctoral student at New York University (NYU), my adviser, Arjun Appadurai, was a gentle captain. A safe harbor. A master of language and an agile thinker, he steadied this ship at just the right moments with a light touch. Our afternoons spent analyzing this field site were some of the most rewarding moments of my graduate career. Craig Calhoun was my anchor. Like an octopus with eight hands constantly at work, he has three hearts. I learned from him the ethic that treats scholarship as a collective endeavor. My accreditation as an ICCAT observer would not have been possible without his founding of the Institute for Public Knowledge (IPK) at NYU. Sally Engle Merry took me under her wing, like an albatross, and modeled how to travel epic distances in an academy still learning how to embrace the contributions of women. She taught me how to be a careful scholar when designing and executing a research project, and did so with extraordinary grace. I learned from her that intuition is an important way of knowing, and curiosity a scholar's greatest motivation. Peder Anker, Nicolas Guilhot, Christine Harrington, Nick Mirzoeff, Anne Rademacher, and Arvind Rajagopal offered critical feedback at key moments in the project's development.

This book has benefited over the years from the generosity of many people who in graduate course work and extended conversation have helped me navigate the horizons of possibility. They include Daniel Aldana Cohen, Philip Alston, Solon Barocas, Franck Billé, Gabi Bockaj, Erin Braatz, Sam Carter, Priya Chandrasekaran, Paul Chevigny, Gabrielle Clark, Jess Coffey, Tine Destrooper, Françoise Dussart, Jan English-Lueck, David Fonseca, David Garland, Radha Hegde, Dale Jamieson, Liz Koslov, Don Kulick, Steven Lukes, Ted Magder, Emily Martin, Samuel Martínez, Joanne Nucho, Matt Powers, Renato Rosaldo, Bambi Schieffelin, Howard Schiffman, Richard Sennett, Mihaela Serban, Anna Skarpelis, Helga Tawil-Souri, Umut Turem, Lesley Turnbull, Johannes Waldmueller, Naomi Waltham-Smith, Richard Wilson, and Robert Wosnitzer.

At Pratt Institute I have accumulated many debts. Chair Gregg Horowitz introduced me to the Department of Social Science and Cultural Studies and mentored me as a junior faculty member learning the ropes. His frequent check-ins when I fell ill modeled how to be present with colleagues. Chair Macarena Gómez-Barris encouraged me to see the potential of this work when I was in the weeds on the tenure track. Her commitment to decolonize and diversify knowledge offers a model necessary to guide us on a more inclusive path. Provost Kirk Pillow and Deans Andrew Barnes and Helio Takai understood the need for intellectual space when I needed it most at the very end. I am certain their support for the book's production will help extend the title's reach. Faculty members May Joseph and Carl Zimring deserve special thanks for helping me navigate academic waters. I remain in solidarity with Sameetah Agha, Cisco Bradley, Ric Brown, Josiah Brownell, Caitlin Cahill, Randy Donovitz, Lisabeth During, Ann Holder, Zat Jamil, Chris Jensen, Luka Lucić, Wendy Muñiz, Darini Nicholas, Tetsu O'Hara, Uzma Rizvi, Mark Rosin, Carolyn Shafer, Ira Stern, Zhivka Valiavicharska, Rebecca Welz, Iván Zatz-Díaz, and the various faculty who have invited me to partake in their studies. Administrators Rosa Cho and Sophia Straker-Babb faithfully do the job of five in two. Librarians Nick Dease, Eric O'Toole, Caroline Skelton, and Holly Wilson were dynamos in the stacks. Nayana Malhotra cleverly designed the book's graphics. The gems known as Pratt students, I hope, have learned as much from me as I have from them. Their creative energy inspired me to break free of academic conventions.

Colleagues at various institutions invited me to present this research while it was still a work in progress. For the generative feedback I received, I thank audiences passing through Columbia University, Dalhousie University, New York University School of Law, Pratt Institute, University of Cambridge, University of Helsinki, University of King's College, and University of Pennsylvania. I benefited from critiques on earlier iterations of this work in conferences of the American Anthropological Association (AAA), the American Ethnological Society (AES), the Association for Environmental Studies and Sciences (AESS), and the Law and Society Association (LSA). In these settings I learned from Nathanael Ali, Megan Bailey, Dean Bavington, Laura Bear, Jennifer Cardinal, Brenda Chalfin, Martha Finnemore, Susanna Fuller, Fiona Haines, Jeffrey Hutchings, Fleur Johns, Benedict Kingsbury, Kaisa Kuurne, Adam Liebman, Laura Ogden, Ian Stewart, Lindy Weilgart, Bethany Wiggin, and Boris Worm, among many others.

Grants from the American Association of University Women (AAUW), the Furthermore Foundation: a program of the J. M. Kaplan Fund, the Horowitz Foundation for Social Policy, the Mellon Foundation, the NYU Council for Media and Culture, the Pratt Faculty

Development Fund, and my university departments supported field research and the craft of writing. A chance encounter with Edvard Hviding in the frenzy of the AAA years ago was, in retrospect, manna from the heavens. Our conversations still fathoms deep have invigorated my passage from the Atlantic to the Pacific, reminding me that oceanic currents are one. His astute vision allowed me to be with this manuscript in its final stages through the grant "Mare Nullius," funded by the Research Council of Norway's TOPPFORSK Programme. I have no conflicts of interest.

The gracious crew of the University of Minnesota Press has masterfully seen this book through to completion. Shelby Connelly, Doug Easton, Emily Hamilton, Eric Lundgren, Maggie Sattler, Heather Skinner, David Thorstad, and Laura Westlund deserve special thanks. Ana Bichanich and the design team developed a stunning cover. Jason Weidemann believed in this project from the very beginning. Steady at the helm. His editorial direction makes clear why transdisciplinary scholarship is fundamental to reimagining ecological futures. The stellar readers he secured were beyond what I could have ever imagined possible. I am humbled to know that the scholars whose work I admire the most took the time to review this manuscript in full. Nayanika Mathur, Daniel Pauly, Anna Tsing, and an anonymous reader elevated the text enormously with their generous feedback and redirection. Friends and colleagues I deeply trust and intellectually respect reviewed chapters in the final push. Matt Canfield, Kari Hensley, Larry Menna, Kim and Ryan Noonan, and Johanna and Louis Römer were lifesavers. Not least, the poetic soul and editorial acumen of Peter Dimock encouraged me to find my voice. He's a writer's writer to the core. His genius for and sensitivity to structural rhythms indelibly mark these pages. Without him, this ship simply would have run aground.

The shortcomings that remain are, of course, my own.

For their unconditional love and support, this book is dedicated to my rock, my lighthouse, my parents, newlyweds at fifty-two years and counting. I can only hope that I inherited the steady, reasoned voice of my pops and the feistiness of my mom to tell it like it is. I have learned from them both a deep respect for the sea. I hope my nieces and nephews will know, as I did, the treasures carrying with them the wonders of the ocean. Uncle Chris, Aunt Martha, my siblings—Pam, Will, and

Tom—my in-laws, my godparents, my cousins have buoyed me in this journey. First mates, physicians, caregivers, and friends far and near have accompanied me through the changing tides of life.

I thank each and every one of you.

For all that, Pa, I am very lucky.

Appendix A

~~~

CONTRACTING PARTIES *to the*
ICCAT CONVENTION, 1967–2012

Contracting Party to the Convention (CPC)	CPC since
Japan	1967
South Africa	1967
United States	1967
Canada	1968
France (in care of Saint-Pierre and Miquelon)	1968
Ghana	1968
Brazil	1969
Morocco	1969
Republic of Korea	1970
Ivory Coast	1972
Angola	1976
Gabon	1977
Russian Federation	1977
Cape Verde	1979
São Tomé and Príncipe	1983
Uruguay	1983
Venezuela	1983
Equatorial Guinea	1987
Republic of Guinea	1991
Libya	1995
United Kingdom (in care of its overseas territories)	1995
People's Republic of China	1996
Croatia	1997
European Union	1997
Tunisia	1997
Panama	1998
Namibia	1999

Contracting Party to the Convention (CPC)	CPC since
Trinidad and Tobago	1999
Barbados	2000
Algeria	2001
Honduras	2001
Iceland	2002
Mexico	2002
Vanuatu	2002
Turkey	2003
Guatemala	2004
Nicaragua	2004
Norway	2004
Philippines	2004
Senegal	2004
Belize	2005
Syria	2005
Saint Vincent and the Grenadines	2006
Egypt	2007
Nigeria	2007
Albania	2008
Mauritania	2008
Sierra Leone	2008
Cooperators	
Chinese Taipei (Taiwan)	1998
Curaçao (formerly the Netherlands Antilles)	2004
Colombia	2009
Suriname	2011
El Salvador	2012

Note: Belize, Panama, and Saint Vincent and the Grenadines were members of panel 2 (Appendix C) but they were not allocated a quota to fish for Atlantic bluefin tuna (Appendix B). Cyprus, France, Italy, Malta, Portugal, Spain, and the UK withdrew from ICCAT as individual CPCs following the accession of the European Community (now the European Union). France and the UK retained membership on behalf of their respective overseas territories not covered by the Treaty of Rome, which established the European Economic Community in 1958. Senegal was an ICCAT Contracting Party from 1971 to 1988, and rejoined in 2004. Not listed are Cuba, which joined as a CPC from 1975 to 1991, and Benin from 1978 to 1994.

Appendix B

~~~

ALLOCATIONS *in* EXPORT QUOTAS *for* ATLANTIC BLUEFIN TUNA

Bluefin in the Western Atlantic

	1982		2011, 2012, 2013	
Total Allowable Catch (TAC), measured in metric tons	1160		1750.00	
Individual Quota Allocation, per Contracting Party (CPC)		% of TAC		% of TAC
Canada	250	21.55	381.66	21.77
France (in respect of Saint Pierre and Miquelon)	n/a		4.00	.23
Japan	305	26.29	301.64	17.24
Mexico	n/a		95.00	5.43
United Kingdom (in respect of overseas territories)	n/a		4.00	.23
United States	605	52.16	923.70	52.78
By-Catch Allowance, as it relates to longline fisheries				
Canada	n/a		15.00	.86
United States	n/a		25.00	1.43

Bluefin in the Eastern Atlantic and Mediterranean Sea

	1999		2010		2011 and 2012		2013	
Total Allowable Catch (TAC), measured in metric tons	32,000.00		19,950.00		12,900.00		13,400.00	
Individual Quota Allocation, per Contracting Party (CPC)		% of TAC		% of TAC		% of TAC		% of TAC
Albania	n/a		50.00	.25	32.33	0.25	33.58	0.25
Algeria	n/a		1012.13	5.07	138.46	1.07	143.83	1.07
China	82	.25	56.86	.29	36.77	0.29	38.19	0.29
Chinese Taipei (Taiwan)	714	2.23	61.48	.31	39.75	0.31	41.29	0.31
Croatia	950	2.97	581.51	2.91	376.01	2.91	390.59	2.91
Egypt	n/a		50.00	.25	64.58	0.50	67.08	0.50
European Union (EU)	20,165	63.02	11,237.59	56.33	7,266.41	56.33	7,548.06	56.33
Iceland	n/a		46.11	.23	29.82	0.23	30.97	0.23
Japan	3,199	10.00	1,696.57	8.50	1,097.03	8.50	1,139.55	8.50

Individual Quota Allocation, per Contracting Party (CPC)	1999		2010		2011 and 2012		2013	
Total Allowable Catch (TAC), measured in metric tons	32,000.00	% of TAC	19,950.00	% of TAC	12,900.00	% of TAC	13,400.00	% of TAC
Korea, Republic of	672	2.10	119.90	.60	77.53	0.60	80.53	0.60
Libya	1,300	4.06	857.33	4.30	902.66	7.00	937.65	7.00
Morocco	820	2.56	1,891.49	9.48	1,223.07	9.48	1,270.47	9.48
Norway	n/a		46.11	.23	29.82	0.23	30.97	0.23
Syria	n/a		50.00	.25	32.33	0.25	33.58	0.25
Tunisia	2,326	7.27	1,573.67	7.89	1,017.56	7.89	1,057.00	7.89
Turkey	n/a		619.28	3.10	535.89	4.15	556.66	4.15

Note: ICCAT member states first instituted export quotas for western and eastern Atlantic bluefin tuna in the years 1982 and 1999, respectively. In 2012, participating EU members in the Atlantic bluefin tuna fishery shared the quota as follows: Cyprus (1%), France (17%), Greece (2%), Italy (31%), Malta (3%), Portugal (5%), and Spain (42%).

Appendix C

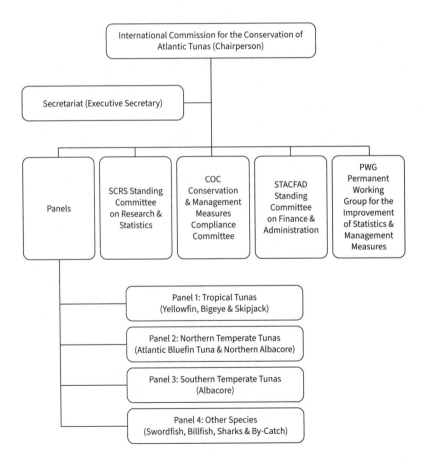

AN ORGANIZATIONAL CHART *of the* INTERNATIONAL
COMMISSION *for the* CONSERVATION *of*
ATLANTIC TUNAS THROUGH 2012

International Commission for the Conservation of Atlantic Tunas (Chairperson)

Secretariat (Executive Secretary)

Panels

SCRS Standing Committee on Research & Statistics

COC Conservation & Management Measures Compliance Committee

STACFAD Standing Committee on Finance & Administration

PWG Permanent Working Group for the Improvement of Statistics & Management Measures

Panel 1: Tropical Tunas
(Yellowfin, Bigeye & Skipjack)

Panel 2: Northern Temperate Tunas
(Atlantic Bluefin Tuna & Northern Albacore)

Panel 3: Southern Temperate Tunas
(Albacore)

Panel 4: Other Species
(Swordfish, Billfish, Sharks & By-Catch)

Notes

〜〜〜

Prologue

1. Simple in approach, trim in form, the tactic I adopt throughout this book to feminize the bluefin in the singular by using the pronouns "she" and "her" is meant to contest and destabilize the commodification of fish. Surely a living being cannot be "it." I forgo the plural, such as "they" and "them," because I try to avoid where possible statistically oriented, aggregate registers of speech such as "species" and "population." I do not want to reduce the bluefin to an abstract "class of being" (Smuts 2001). The feminized, singular pronoun is not meant to conjure up "Mother Nature" per se—although it could—but to remind the reader that nonhuman natures under extractive capitalism, like women's work, have long been kept off the balance sheet, their worth expunged from cost-benefit calculi. The use of "she" and "her" is not meant at all to reinforce what I call the savior plot in chapter 3, as if the bluefin is a damsel in distress, but to instigate an appeal to respect a fellow living being. Sylvia Wynter reminds us that "scholars necessarily function as the *grammarians* of our order; that is, as 'men and women' who are well-versed" in how to arrange a body of facts within a framework consistent with their value system in the society in which they belong (1994, 55; emphasis in original).
2. Malamud 1998.
3. Collette, Reeb, and Block 2001, 1.
4. Safina 1997, 59. The ecologist Carl Safina offers a brilliant discussion of the bluefin's speed, drag, and agility here.
5. Block et al. 2001, 1313.
6. Ellis 2008, 42–43.
7. Ritvo 1997.
8. Block quoted in Ellis 2008, 37.
9. Issenberg 2007.
10. Roberts 2007, 280.
11. Corson 2007, 249.
12. Bestor 2004, 309.
13. Miyake et al. 2010, 63.
14. Kopytoff 1986, 90.
15. Myers and Worm 2003.

16. As of this writing, the International Union for Conservation of Nature (IUCN) lists the Pacific bluefin tuna as "vulnerable" and the Southern bluefin tuna as "critically endangered." These designations are based on conservative estimates of decline: http://www.iucnredlist.org/.
17. Hardin 1968.
18. Longo, Clausen, and Clark 2015.
19. These figures are based on a summary report by the Principality of Monaco issued in 2009 in an effort to list Atlantic bluefin tuna as "endangered" under the Convention on International Trade in Endangered Species of Wild Fauna and Flora (CITES). The report relies on the "stock assessment" for Atlantic bluefin tuna conducted in 2008 by ICCAT's scientific advisory committee.
20. See the report by the International Consortium of Investigative Journalists (ICIJ), "Looting the Seas: How Overfishing, Fraud and Negligence Plundered the Majestic Bluefin Tuna" (2010), partly funded by the Pew Environment Group, at https://www.ire.org/resource-center/stories/24880/.
21. Maggio 2000, 9.
22. Longo and Clark 2012, 208.
23. Maggio 2000, 35.
24. Fonteneau 2008, 4.
25. Fromentin and Powers 2005, 285.
26. Fisher quoted in Longo and Clark 2012, 217.
27. Telesca 2017, 2018.
28. Bestor 2000, 59.
29. Roberts 2007, 197.
30. Whynott 1995, 105.
31. Ibid., 1995, 50.
32. Maggio 2000, 139.
33. Hemingway quoted in ibid., 139–40; emphasis added.
34. Haraway 2008.
35. Appadurai 1986.
36. Kopytoff 1986, 73.
37. McNeill and Engelke 2014 write: "The global marine fish catch quintupled between 1950 and 2008, while [human] population growth nearly tripled. As a rough estimate, one can say that 60 percent of the expansion of the marine fish catch derived from population growth." In other words, about 40 percent of the expansion can be attributed to the intensity of the fishing effort (52–53). Subsidies for fleets from industrialized countries and investments in more "efficient" technology contributed to rapid fish decline, as detailed in Finley 2017.
38. Telesca 2018.
39. Carson 1989 (1950), 93.

Introduction

1. Issenberg 2007.
2. Graeber 2001, 1–2.

3. Telesca 2017. See Epstein 2008 for a discussion of language in ocean gover-nance more generally.
4. Appadurai 1986.
5. Clarke and Haraway 2018 offer a discussion not of "population" but of kin in policy making to forge connections across generations.
6. Telesca 2017.
7. See Ceballos et al. 2015. Lay audiences should consult the Pulitzer Prize–winning Kolbert 2014. Because extinction is not only a biological event, but a social and cultural phenomenon rooted in political economy, the following literature from the social sciences and humanities is informative: Dawson 2016; Grusin 2018; Heise 2016; Parreñas 2018.
8. The leading scientists of extinction, Gerardo Ceballos and Paul R. Ehrlich write: "Mass extinctions are defined as the loss of the majority of species in a relatively short geological time, caused by a catastrophic natural event. Some scientists argue that there is no reason for concern about the sixth mass extinction because extinction is normal, simply an inevitable consequence of the process of evolution. This misunderstanding ignores some critical issues. First, the rate of species extinction is now as much as 100 times that of the 'normal rate' throughout geological time. Second, like the past mass extinctions, the current episode is not an inevitable consequence of the process of evolution. Rather, it is the result of a rare event changing the environment so quickly that many organisms cannot evolve in response to it" (2018, 1080). What that "rare event" is and how to define its scope are hotly debated, evident in the burgeoning discussion about the Anthropocene.
9. Tsing et al 2017, G4. A landmark report from the Intergovernmental Science-Policy Platform on Biodiversity and Ecosystem Services (IPBES) issued in May 2019 exposes the enormity of the crisis. In the most comprehensive assessment of biodiversity up to this date, prepared by more than 150 experts from fifty countries, with additional contributions from a further 250 representatives from the natural and social sciences, the report drew on nearly fifteen thousand references, including peer-reviewed scientific papers and information from governments. According to the IPBES Web site, "[The report] is also the first global assessment ever to systematically examine and include indigenous and local knowledge, issues and priorities." Its conclusion: one million animal and plant species are threatened with extinction.

To assist policy makers in decision making, the IPBES report also ranked, "the five direct drivers of change in nature with the largest relative global impacts so far. These culprits are, in descending order: (1) changes in land and sea use; (2) direct exploitation of organisms; (3) climate change; (4) pollution and (5) invasive alien species." The organized extraction of the bluefin by ICCAT member states corresponds with one of the top drivers of extinction. https://www.ipbes.net/news/ipbes-global-assessment-preview. Other reports by UN agencies paint a similarly grim picture of the future of marine life—if there is no change in course. See, for example, the "Special Report on the Ocean and Cryosphere in a Changing Climate" issued by the

Intergovernmental Panel on Climate Change (IPCC) in September 2019. It is available online at https://www.ipcc.ch/srocc/home/.

10. As of this writing, the start date for the Anthropocene is far from settled. Some geologists mark the threshold between the Holocene and the Anthropocene at the rise of the Industrial Revolution in 1800. Indigenous peoples and descendants of slaves still combatting the legacy of the colonist and the master of the plantation counter that a more appropriate beginning is 1492, when Columbus unleashed his brutal campaign to "discover"—and exploit— the Americas. Simon L. Lewis and Mark A. Maslin (2015) offer evidence of an "Orbis Spike" in 1610, when the "Little Ice Age" emerged as an atmospheric response to the carnage wrought by colonialism. Still others favor the start date of 1945 when the U.S. military detonated the first atomic bomb in the state of New Mexico. Feminist scholars contest who "man" in "anthropos" is, while others worry that the Anthropocene flattens responsibility for the climate crisis by lumping the rich with the poor under the universalizing banner "man" or "human." Likewise, some writers think the Anthropocene label is not useful at all and actively write against it, often by offering new conceptualizations such as Capitalocene, Plantationocene, and Chthulucene. See Bonneuil and Fressoz 2017; Chakrabarty 2009, 2017; Demos 2017; DiChiro 2016; Grusin 2017; Haraway 2016; Lorimer 2015; Malm and Hornborg 2014; Moore 2015; Purdy 2015; Tsing et al. 2017; Yusoff 2018.

11. Steffen et al. 2011. McNeill and Engelke 2014 write: "The Great Acceleration in its present form cannot last for long." The planet is finite. "But the Anthropocene, barring catastrophe, is set to continue" (5). I worry that the Great Acceleration is too innocuous a phrase. Scrutiny of this naming practice deserves full attention elsewhere.

12. For graphic depictions of these trends, see http://www.anthropocene.info/great-acceleration.php.

13. Pauly 2010, 63–91 documents that the global capture of wild fish peaked in the mid-1990s. Decline in catch thereafter does not imply regulatory restraint.

14. Holt 2006.

15. DeSombre and Barkin 2011, 95.

16. Maggio 2000, 140.

17. Smuts 2001.

18. Williams 1977.

19. Finance capital is not only shorthand for the mechanism bankrolling the accumulation of wealth today. Finance capital also provides the very logics, protocols, and cultural representations by which risk management operates— including in fisheries. For discussion on the derivative as the primary instrument to generate more, and more, wealth, see Appadurai 2016; Borch and Wosnitzer forthcoming; LiPuma and Lee 2004; Martin 2009; Wosnitzer 2016. I touch on these ideas in Telesca forthcoming.

20. Rosaldo 1989.

21. Bailey 2013; Bavington 2010; Bonanno and Constance 1996; Greenberg 2010;

Knecht 2006; Kurlansky 1998; Pauly 1995, 2009; Probyn 2016; Rigney 2012; Safina 1993, 1997.

22. Bonneuil and Fressoz 2017.

23. Rosaldo 1989, 110.

24. Myers and Worm 2003.

25. The term "value" has a long, thorny history in social theory. Its literature is massive. Marxian perspectives alone are many. Harvey 2006 (1982) is prime among them. Some from anthropology and sociology include Appadurai 1986; Camaroff and Camaroff 2005; Graeber 2001; Hitlin and Piliavin 2004; Spates 1983. Massumi 2018 considers revaluation central to the "postcapitalist" project. This book does not address how value is created. Instead it tries to capture the lethal *effects* of a hegemonic regime of value rooted in the commodification of living beings.

26. Kopytoff 1986, 90.

27. Bekoff 2004 offers a perspective from ecology and evolutionary biology.

28. French 1999, 172.

29. Ferguson 1988, 493.

30. Kopytoff 1986, 73.

31. Appadurai 1986, 13.

32. Ibid., 15. See Ballestero 2019 for discussion of water as a human right and commodity.

33. Moore 1978, 11.

34. Kopytoff 1986, 83.

35. Shukin 2009, 11. See also Helmreich 2008.

36. Shukin 2009, 15–16; emphasis in original.

37. Haraway 2008, 19–27 critiques Jacques Derrida precisely for his inability to take seriously the lifeworld of the "animal" and thus to leave unacknowledged all the ways in which his own lifeworld could be enriched by, say, his cat.

38. Shukin 2009, 17. See also Sullivan 2017. The treatise on "natural capitalism" is found in Hawkin, Lovens, and Lovens 1999.

39. Foucault 2008.

40. Hart and Negri 2000, 27.

41. Ingold 2012, 435. Ingold clarifies the distinction between studies of "the ecology of materials" and those of the more mainstream "material culture." This book is sympathetic to the former project. It favors Shukin's formulation of "animal capital" because it accommodates and addresses the ecology of organisms (both human and nonhuman) bound up in webs of life.

42. Bolster 2012; Roberts 2007.

43. Holm 2012.

44. Cullis-Suzuki and Pauly 2010.

45. Finley 2017.

46. https://www.iccat.int/Documents/Commission/BasicTexts.pdf.

47. See also a story on "trade threats—reportedly related to the valuable banana industry" that influenced the decisions of the ICCAT commission in 2008 here:

https://www.ictsd.org/bridges-news/biores/news/iccat-tuna-quota-reductions -not-enough-environmentalists.

48. https://www.reuters.com/article/us-europe-migrants-idUSKBN0U50WI201 51222.

49. Koskenniemi and Lehto 1996, 554.

50. Singh-Renton 2010, 13.

51. Sloan 2003.

52. Pauly 2009.

53. Franklin 2006.

54. Geertz 1978, 29.

55. This book does not intend to endorse the project of "one-worlding" (de la Cadena and Blaser 2018). In fact, I hope this book is read as a critique of the one-worlding project, seen from the inside, as global elites seek to order and manufacture the objectivity and universality of their claims, accomplished through law and fisheries science, which concentrates anthropocentric property relations in the hands of the rich member states that participate in the ICCAT apparatus.

56. Gilroy 1993, 55.

57. Olds and Thrift 2005, 270.

58. "Global governance" has many expressions. Terms such as "global administrative law," "global regulatory system," "international regime," and "international organization" are highly contested. As regulatory bodies operating outside of or alongside domestic law proliferate, so too does the drive to name those practices. Scholars who study or construct these terms participate in a normative project, not simply in a "taxonomical exercise" or in "the promulgation of practical technical solutions to well-defined and accepted problems," write Kingsbury, Krisch, and Stewart 2005, 42. I recognize that some technocrats conceptualize "regional and sectorial bodies" represented by regional fisheries management organizations (RFMOs) separately. It is beyond the scope of this book to settle the ground of these classifications.

59. Nappi 1979, 178.

60. Halliday and Carruthers 2009; Martello and Jasanoff 2004. Core anthropological treatments of globalization include Appadurai 1996; Ong and Collier 2005; Tsing 2005.

61. Kingsbury, Kirsch, and Stewart 2005.

62. Cassese 2005.

63. In ocean governance the most well known exception is the International Whaling Commission (IWC), which formed in 1946. Cooperation among fishing nations developed as early as 1882 in the Convention on North Sea Overfishing, write DeSombre and Barkin 2011, 84.

64. Cassese 2005.

65. http://www.fao.org/fishery/rfb/search/en.

66. The five tuna regional fisheries management organizations (RFMOs) include: ICCAT; the Inter-American Tropical Tuna Commission (IATTC), which has

enjoyed jurisdiction over the eastern Pacific since 1948; the Commission for the Conservation of Southern Bluefin Tuna (CCSBT), which formed in 1994; the Indian Ocean Tuna Commission (IOTC), which formed in 1996; and the Western and Central Pacific Fisheries Commission (WCPFC), which formed in 2004. These tuna RFMOs regulate the catch of pelagic creatures in the open ocean beyond the jurisdiction of any one nation-state.

67. Cassese 2005, 671–72.
68. Benvenisti and Downs 2007; Porter 1995.
69. Rosenau 1992, 4.
70. Slaughter 2004, 5.
71. Chayes and Chayes 1995; Keohane and Nye 2001b.
72. Chayes and Chayes 1995.
73. Polanyi 2001 (1944), 16.
74. Boltanski and Chiapello 2005, 4; emphasis in original.
75. Moore 1978, 21.
76. Block 2001, xxvi.
77. Scott 1998.
78. Williams 1977.
79. Barnett and Finnemore 2004 offer useful ways to theorize authority and autonomy as they relate to international organizations.
80. In the U.S. example, Slaughter 2004, 36–37 documents that financing the foreign-affairs budgets to participate in regimes such as ICCAT has "increased dramatically across the board, even as the State Department's budget has shrunk."
81. Bilateral agreements compete with the ICCAT convention. They are one of the great elephants in ICCAT meeting rooms.
82. Hardt and Negri 2000, 15.
83. ICCAT member states may grant "Cooperating Status" to a "Non-Contracting Party, Entity or Fishing Entity." These states fish in the Atlantic and its nearby seas but do not enjoy full ICCAT membership. Although they may claim catch quotas, their membership is reviewed annually and they cannot vote (unlike Contracting Parties). See Appendix A for details.
84. ICCAT member states have agreed to institute "capacity-building" measures in such areas as scientific research and statistical training, technical assistance in the fulfillment of reporting obligations, and financial aid for travel to meetings. The rich trading areas of the European Union, Japan, and the United States pay for them. ICCAT member states may also agree to dip into the capital fund. Some delegates from developing coastal states expressed to me that they felt these programs came with strings attached.
85. Callon 1998, 7.
86. Geertz 1978, 29.
87. Slaughter 2004.
88. Wedel 2009.
89. Although environmental campaigns have drawn attention to the sharp decline

in sharks, whether ICCAT member states categorized sharks (and billfish) as "by-catch" or as targeted species for market was hotly contested while I was in the field.

90. Harrington, Myers, and Rosenberg 2005.

91. Small tunas include blackfin tuna, bullet tuna, Atlantic bonito, plain bonito, Serra Spanish mackerel, frigate tuna, king mackerel, little tunny, West African Spanish mackerel, Atlantic Spanish mackerel, wahoo, and dolphinfish (see www.fishbase.org). Because these creatures were often misidentified at port, because biological and statistical information about them was limited, and because large industrial fleets discarded them at sea or sold them to local markets—which implied that they bypassed global trading networks entirely—regulating them fell to other bodies, not to ICCAT. Although IC-CAT narrowly regulates the quantity or "abundance" of some, but not all, pelagic commercial fish in the Atlantic basin, it does not regulate the health of sea creatures or the quality of fish meat. Troubles associated with high levels of mercury, plastic, and other toxins accumulating in the bodies of fish are not ICCAT's to solve. These are treated, if at all, as variables among others at scientific meetings in investigations about fish mortality, discussed in chapter 4.

92. http://www.iccat.int/Documents/SCRS/ExecSum/SMT_EN.pdf.

93. Telesca 2017.

94. Chayes and Chayes 1995.

95. The first trade sanction came in ICCAT Recommendation 96–11. Not all trade sanctions remain active once adopted, because ICCAT member states review each year whether to maintain or drop the "identification." The creature provoking the creation of trade sanctions at ICCAT was, predictably, Atlantic bluefin tuna.

96. Chayes and Chayes 1995; Keohane and Nye 1974; Slaughter 2004.

97. Keohane and Nye 1974.

98. Billé 2016.

99. Moore 1978, 36.

100. Deloughrey 2019.

101. Feldman 2011, 33.

102. Marcus 1995.

103. Feldman 2011, 32.

104. Kirksey and Helmreich 2010.

105. Haraway 2008.

106. See, for example, Bear and Mathur 2015; Mathur 2016.

107. Merry 2006; Riles 2011; Wilson 2001, 2011 offer models.

108. Shore and Wright 1997; Shore, Wright, and Però 2011.

109. Mathur 2016, 176.

110. Ferguson 1990; Halliday and Carruthers 2009; Neumann 2012, 84.

111. Feldman 2011, 33.

112. Appadurai 1986, 48.

113. Dittmer 2017.

114. According to ICCAT document 05–12 titled "Guidelines and Criteria for Granting Observer Status at ICCAT Meetings," observer applications are accepted unless one-third of Contracting Parties object to an observer's participation in writing. The application must first pass the secretariat. I heard in my travels through the ICCAT network that at least one researcher was denied access to ICCAT meetings.

115. Verdery 2018, 18–19 discusses the implications of surveillance on ethnography: "mistrust and doubt eats away" at relations while in the field. This dimension of ethnographic research with diplomats deserves full treatment elsewhere.

116. "State" is as thin and imprecise as "industry" and "civil society." There is variability within all these categories. They can vary by delegation and by year. Such labels are imperfect and are meant to designate a broad category of participation in the ICCAT regime. I do not intend to use them in my analysis as ideal types.

117. Marks 2003, 461.

118. Koskenniemi 2006, 608–9.

119. I use the terms "expert," "bureaucrat," "technocrat," and "diplomat" somewhat interchangeably throughout this book to reference ICCAT delegates. Sometimes these terms are not interchangeable, as Iver Neumann points out. Designation is context-specific. In fact, there are moments when the statesperson or politician is not the diplomat at all: although steeped in politics, the diplomat aims to develop an "institutional voice" to maintain the system while the politician seeks to gather support from others to change the system (Neumann 2012, 86–87). This tension explains, in part, ICCAT's glacial pace of change.

120. To slide between *the personal* (not representing an official state or NGO policy, program, and position), *the official* (representing an adopted state or NGO policy, program, and position) and *the private* (outside of or beyond diplomatic affairs) is how diplomatic work itself is done, writes Neumann 2012, 126–27.

121. Holmes and Marcus 2005, 248.

122. Dittmer 2017, 19.

123. Webster 2008 offers an analysis of ICCAT to predict a member state's negotiating position based on the author's "vulnerability response framework."

124. I embraced feminist methodologies, and took inspiration from such authors as Haraway 1988, Joseph 2019, and Reinharz 1992.

125. Rosaldo 1989, 120–21.

126. Scheper-Hughes 1995, 419 offers guidance on "personally engaged and politically committed" ethnography, as does Hughes 2017.

127. Williams 1977.

128. Recent efforts to amend the ICCAT convention briefly illustrate the point.

129. Few fishers participated in ICCAT meetings. The fishers who did participate overwhelmingly represented the industrial or commercial sector, not the recreational one.

130. Hamblyn 2009.
131. Whether "ecosystem-based management" will solve the overfishing problem by "managing" the ecosystem as a whole rather than by individual fish remains to be seen. This book claims that these kinds of approaches must confront and dislodge the prevailing predatory regime of value if they are to offer a meaningful path forward.
132. Appadurai 1986.
133. Haraway 2016, 7.
134. Varoufakis 2017.

1. A History of the Bluefin Tuna Trade

1. Kingsbury, Krisch, and Stewart 2005.
2. For a discussion of the functions of international regulatory agreements, see Chayes and Chayes 1995, 274–82.
3. Issenberg 2007, 233.
4. Tsing 2015.
5. I deeply thank Arjun Appadurai for helping me to distill the phrase "commodity empire."
6. Feldman 2011, 46.
7. Ghosh 2016, 146. From a geological perspective, recent studies of the "Little Ice Age" suggest that European colonization cooled the climate in the aftermath of killing 56 million indigenous people in the Americas by 1610 (Lewis and Maslin 2015). For perspectives on indigenous resistance and the refusal to submit to extractive capitalism, see Gómez-Barris 2017.
8. My sincere thanks to Anna Tsing for suggesting that this history begin not with people but with bluefin. Her own essay (2012) provides a model daring to tell a feminist history of entanglements in the *longue durée* through the lens of a mushroom.
9. Bolster 2012, 6–7.
10. Ibid., 19–20.
11. Graham and Dickson 2001, 150–52.
12. Block et al. 2001, 1310.
13. https://www.science20.com/news_articles/bluefin_tuna_hearts_are_tougher_than_yours_how_they_stay_warm_in_the_cold-153027.
14. Mather, Mason, and Jones 1995, 114.
15. Block et al. 2001, 1313.
16. Innis 1940 offers insight on imperialism through the cod fishery.
17. Myers 2001.
18. Benton, Edelstein, and the Editorial Collective 2011.
19. Hardt and Negri 2000, 3.
20. Anderson 2002 (1983), 7; emphasis added. Nationalism as affect was uneven and not always expressed as solidarity at ICCAT. Sloan 2003 details the palpable tensions within the U.S. delegation.
21. Beckert 2014; Innis 1940; Mintz 1985.
22. Schmitt 1997 (1942). See Connery 2001.

23. Mancke 1999, 233.

24. Cowen and Smith 2009, 24–25. Cowen and Smith's formulation of geoeconomics differs from others, including that of Edward Luttwak, who is credited with coining the term "geoeconomics" in 1990 despite the fact that the first formulation may have been by the geographer Jacques Baudeville in 1966. Cowen and Smith rightly critique Luttwak for assuming that globalization results from "the natural evolution in economic affairs" and that the geoeconomic transition is even and complete (38). The analysis put forward here also distances itself from Luttwak's formulation (Luttwak 1990).

25. Although some readers may understand what I am describing here as "neoliberalism," I favor geoeconomics as an organizing framework in the ICCAT context because it does not emphasize a temporal phenomenon at the expense of what is in reality both a temporal *and* a spatial one. The reordering of the ocean into pockets of mobile property within and beyond exclusive economic zones (EEZs), detailed in chapter 2, makes this point clear.

26. Mitchell 1998, 84.

27. Escobar 1993, 4.

28. Mitchell 1998, 89.

29. Ibid., 90.

30. Cooper 2005, 194. See Hardt and Negri 2000, xiv–xv for discussion on the limits of empire as metaphor.

31. Hardt and Negri 2000, xii; emphasis in original.

32. Foucault 2007b, 367; Gordon 1991, 4–5. Patton 2016 offers an excellent overview of biopower and Nealon 2016 how it relates to animal studies.

33. Foucault 2007b, 248.

34. Hardt and Negri 2000, xv; emphasis added.

35. Some readers may associate this analysis with what Foucault calls "governmentality," or the "art of government," the "conduct of conduct" (Gordon 1991, 48)—that is, the practices, activities, programs, campaigns, documentations, and other "technologies" that take as their goal, in this context, the regulation of fish and the people that catch them, who together are guided by an authority that orchestrates, directs, and is responsible for what people do and what happens to them when supplying the market with bluefin tuna. Rather than liberating subjects, knowledge enables the continuous, pervasive control over fish and the way people conduct themselves, which in turn is so productive that it continues down a vortex of further knowledge, inquiry, and disclosure. Arun Agrawal (2005) calls this condition "environmentality." This mode of analysis is very helpful. All these elements are in play here. But to fully appreciate the dynamics of ICCAT's predatory regime of value, to observe life beyond and against ICCAT's overrationalizations, to refute the blatant absurdities of ocean governance, to foreground the play element of regulatory action, other orientations are also needed.

36. Curtis 1988. My cursory treatment of the history of empire traced though Atlantic bluefin tuna is not intended to be complete or exhaustive. I do not intend to impose an arbitrary continuity or a single coherent linear narrative.

I cite history from the Mediterranean basin over other Atlantic regions because, simply, the written record about the tuna's capture is extensive there.

37. Appadurai 2013; Shukin 2009.
38. Tsing 2015, 111.
39. Longo and Clark 2012, 208.
40. Maggio 2000, 55; Brown 1968, 47.
41. Bekker-Nielsen 2005.
42. Stolba 2005.
43. Grusin 2017.
44. Aristotle quoted in Brown 1968, 56.
45. Lytle 2006.
46. Brown 1968, 59. Some classicists claim that Byzantium may be "accorded undue prominence in our ancient evidence on the export of food from the Black Sea" and that much of the fish probably "came not from but via Byzantium" (Braund 1995, 162).
47. Horden and Purcell 2000, 194.
48. Sara 1980, 131. Purcell 2003, 341–43 offers a discussion of how Greeks and Romans differed in their perception of foodways. Fish sauce (*garum*) was critical to their diets.
49. Farwell 2001, 391.
50. Plutarch quoted in Maggio 2000, 56.
51. Fagan 2006, 6.
52. Horden and Purcell 2000, 194.
53. Ibid., 195.
54. Purcell 1995, 147.
55. Sara 1980, 130; Longo and Clark 2012, 208.
56. Sara 1980.
57. Longo and Clark 2012, 201.
58. Ibid., 210.
59. Fagan 2006.
60. Brown 1968, 58.
61. Longo and Clark 2012, 212. Di Natale 2012 describes the first visual evidence of traps as they are more or less used today: from George Braun and Franz Hogenberg in their work, *Civitates Orbis Terrarum* (1572–98), which showed images of the Spanish *almadraba* in etchings by Georg Hoefnagel from Cadiz (1572) and Conil (1575).
62. Fagan 2006.
63. Carruthers and Espeland 1991; Poovey 1998; Weber 1978 (1922).
64. Telesca 2017.
65. Longo and Clark 2012, 212.
66. Mather, Mason, and Jones 1995, 1.
67. Grandin 2014, 8.
68. Baptist 2016, 35, 39.
69. Cassady 1974.
70. Longo and Clark 2012, 213–14.

71. Roberts 2007, 281.
72. Bolster 2012.
73. Mather, Mason, and Jones 1995, 17.
74. Cullis-Suzuki and Pauly 2010, 1036.
75. Roberts 2007, 189.
76. Suárez de Vivero and Rodríguez Mateos 2002.
77. Borgström 1964, 25.
78. Ibid., 29.
79. Roberts 2007, 281.
80. The April 2010 report "Fiscal and Monetary Policies of Japan in Reconstruction and High Growth (1945–1971)," issued by the Institute of Fiscal and Monetary Policy of the Ministry of Finance, Japan, is available at http://www.mof.go.jp/english/pri/publication/policy_1945-1971/index.htm.
81. Johnston 1953.
82. Bestor 2004, 116–17.
83. Issenberg 2007, 241.
84. Whynott 1995, 83.
85. Bergström 1964, 198.
86. Ibid., 272–74.
87. Japan is not alone in this regard. According to the FAO, "Since 2002, China is the world's largest exporter of fish and fishery products, with exports reaching USD19.7 billion in 2015. In the last few years, China has significantly increased its imports of fish and fishery products and became the world's third largest fish importer since 2012, with imports worth USD8.5 billion in 2014." I heard in my travels through the ICCAT network that China was more concerned about securing great quantities of fish for its billion people than it was about the niche market for bluefin tuna. See http://www.fao.org/fishery/facp/CHN/en.
88. Holm 2012, 72–73.
89. Ibid., 78.
90. Hardt and Negri 2000, 40.
91. Foucault 2007b, 33. This passage focuses not on fish but on grain shortages.
92. Telesca 2018.
93. Telesca 2017.
94. Finley 2011, 2. See also Smith 1994.
95. Daniel and Minot 1961 (1954), 161–62; emphasis added.
96. Huxley quoted in Cushing 1988, 117.
97. Finley 2011, 2.
98. Larkin 1977, 1–2.
99. Ibid., 6.
100. Nibert 2002.
101. Mitchell 1998, 86.
102. Holling 1973, 1.
103. Mitchell 1998, 85–86.
104. Kingsland 1985, 4.

105. Sinclair and Solemdal 1988.
106. Finley 2011; Smith 1994.
107. Finley 2011, 136.
108. Ibid., 9.
109. Hubbard 2006, 189.
110. Finley 2011, 88.
111. Finley 2011.
112. Holling 1973, 2; emphasis added.
113. Larkin 1977.
114. Finley 2011, 3–4.
115. Chapman quoted in ibid., 155.
116. Bonneuil and Fressoz 2017, 22–23.
117. Levinson 2006; Vine 2009, 18.
118. McNeill and Engelke 2014, 130.
119. Fonteneau 2008, 6.
120. International Commission for the Conservation of Atlantic Tunas 2009, 5.
121. Fonteneau 2008, 6.
122. Ibid., 8.
123. International Commission for the Conservation of Atlantic Tunas 2009, 5.
124. The FAO was important to ICCAT's formation, above all else because it is the UN agency with the legal competence over world fisheries. Although the FAO initiated, developed, and implemented the Atlantic tuna commission, ICCAT is administered independently from the FAO. This arrangement differs from the direct control the FAO enjoys over other tuna RFMOs by virtue of its Article XIV, such as over IOTC.
125. The seventeen countries formally participating in the meeting that finalized the ICCAT convention were Argentina, Brazil, Canada, Cuba, the Democratic Republic of the Congo, France, Japan, Portugal, the Republic of Korea, the Republic of South Africa, Senegal, Spain, the Union of Soviet Socialist Republics, the United Kingdom of Great Britain and Northern Ireland, the United States of America, Uruguay, and Venezuela. Observers included the Federal Republic of Germany, Italy, and Poland.
126. https://www.iccat.int/Documents/Commission/BasicTexts.pdf; emphasis added.
127. Fonteneau 2008, 10.
128. Suárez de Vivero and Rodríguez Mateos 2002, 143.
129. Ibid., 146.
130. Pagden 1994, 2.
131. http://www.fao.org/in-action/globefish/fishery-information/resource-detail/en/c/338172/.
132. Roberts 2007, 280; Issenberg 2007.
133. Longo and Clark 2012, 216.
134. Anonymous fisher quoted in ibid.
135. Miyake et al. 2010, 63.
136. Issenberg 2007, 75.
137. Corson 2007, 145.

138. Borgström 1964, 55.
139. Issenberg 2007, 75.
140. Wholesaler quoted in ibid., 76.
141. Appadurai 2013, 38.
142. Bestor 2004, 118. To actualize this policy—which implied the transition from a command to a market economy in the restructuring of postwar Japan after the loss of its colonies—the Japanese government issued subsidies and loans from the newly formed Reconstruction Finance Bank (RFB) mediated to some degree by the Bank of Japan. It used these financial instruments as operating funds and for capital investment in coal, chemical fertilizers, oil, and the like. Note that 45 percent of the total finances serving the pelagic whaling industry came from the RFB. By 1948, the RFB funded 64 percent of the finances serving the five biggest fishing companies (Endo and Yamao 2007, 169–70).
143. Although Japan remains the largest importer of bluefin tuna today, sushi and sashimi consumption is declining there as consumers gradually change their preferences toward lower-cost foods. The amount of whole tuna auctioned at Tokyo's famed Tsukiji marketplace has waned now that supermarkets, restaurant chains, and other retailers buy 70–80 percent of frozen or chilled fish outside the auction system. At the same time, other countries such as Italy, Spain, and the United States have increased their bluefin consumption: http://www.fao.org/in-action/globefish/fishery-information/resource-detail/en/c/880744/.
144. Issenberg 2007, 3–4; 39.
145. Ibid., 2.
146. MacAlpine quoted in ibid.
147. Okazaki quoted in ibid., 5.
148. Ibid., 6.
149. Bestor 2003, 304.
150. Issenberg 2007, 13.
151. Ibid., 36. Not to be undercut by their Canadian counterparts, American fishers, led by Atlantic Coast Fisheries in Sandwich on Cape Cod, sent chilled bluefin tuna to Japanese fish markets. Their success was such that Atlantic bluefin tuna from the American and Canadian Maritimes are today collectively called "Boston bluefin" at Tsukiji. It is the best bluefin tuna money can buy.
152. LiPuma and Lee 2004, 71.
153. Ibid., 74.
154. Tsing 2015, 110.
155. See page 247 in the 2010 report by the Japanese Ministry of Finance titled "Fiscal and Monetary Policies of Japan in Reconstruction and High-Growth, 1945 to 1971," available at http://www.mof.go.jp/english/pri/publication/policy _1945-1971/Part2-Chapter6.pdf.
156. Whynott 1995, 87–88.
157. Issenberg 2007, 169–72.

158. Whynott 1995, 89.

159. Issenberg 2007, 168–69.

160. Mitchell 2006, 1118.

161. To be clear, Levinson (2006) explores ship (not air) box containers, which, once released from port, travel by train and eighteen-wheeler to deliver goods across the land. According to the International Chamber of Shipping, "The international shipping industry is responsible for the carriage of around 90% of world trade." Nicole Starosielski (2015) documents that 99 percent of all Internet traffic runs by undersea cables. These figures illustrate the extent to which the infrastructure for the global economy runs on the back of the ocean, a topic that deserves full analysis elsewhere. See http://www.ics-ship ping.org/shipping-facts/shipping-and-world-trade.

162. Cowen and Smith 2009, 42.

163. Moore 1978, 8.

164. Big thanks to Larry Menna for helping me to see this important distinction.

165. Hardt and Negri 2000, 9.

166. Telesca forthcoming.

167. Gambs 1946, 92.

2. A "Stock" Splits

1. According to the Atlantic Tunas Convention Act of 1975, the U.S. Advisory Committee to ICCAT is composed of three commissioners representing the federal government, the industrial sector, and the recreational sector.

2. Tsing 2015, 127.

3. Telesca 2017.

4. Tsing 2015, 133.

5. Appadurai 1986.

6. Chayes and Chayes 1995.

7. Five nation-states fishing in the Pacific—Costa Rica, Ecuador, Mexico, Panama, and the United States—adopted catch limits for yellowfin tuna through the Inter-American Tropical Tuna Commission (IATTC) in 1966, thereby establishing the first international measure to regulate the catch of high-seas tunas. The quotas for Atlantic bluefin tuna signaled an entirely new era in ocean governance at the dawn of UNCLOS III.

8. Benton, Edelstein, and the Editorial Collective 2011.

9. Appadurai 2013.

10. Longo, Clausen, and Clark 2015.

11. Law and legality are not ready-made, self-evident categories of understanding. This book adopts the analytic approach put forward by critical legal theorists in the 1980s. In this view, law and politics are not separate, hermetically sealed spheres in which law safeguards decision makers against politics, as a liberal theory of politics would have it (Koskenniemi 2006, 5). Law and society are not two separate realities that either correspond to each other or appear out of sync. Law is not a passive reflection of greater political and economic forces. Neither is it a force that can be used independently as a

blunt instrument to compel lasting economic and political change (see Tamanaha 2006). These perspectives rest on the assumption that law, economy, and state can be distinguished as separate, autonomous fields in practice that "create simple networks of causal relation" (Koskenniemi and Lehto 1996, 533). By adopting a sociolegal approach, this book does not bog itself down in a classificatory debate about where to drawn the line in what constitutes "the law." Turem and Ballestero 2014 offer guidance. I thank Matt Canfield for encouraging me to clarify this point.

12. Turk 1976, 186; emphasis in original.

13. De Sousa Santos 1987, 281; see also Geertz 1983.

14. Turner 1967.

15. Neumann 2012, 181–82; emphasis added. See also Dittmer 2017, 14.

16. Koskenniemi 2006, 590.

17. Alan Harding quoted in Moore 1978, 9.

18. Marks 2009, 2.

19. Moore 1978, 49–53.

20. Ibid., 39.

21. Webster 2008.

22. Koskenniemi 2006, 591.

23. Dezalay and Garth 1995, 36.

24. Turner 1967.

25. Koskenniemi 2006, 610.

26. Benvenisti and Downs 2007.

27. Moore 1978, 26–27.

28. The legal loophole of flags of convenience is a short but incomplete answer.

29. Chayes and Chayes 1995, 27. Kingsbury 1998 offers an overview of different conceptions of compliance in international law. Ali 2018 discusses different modalities of global governance in the context of counterterrorism. Anthropological treatments of sovereignty are many. The ones informing this analysis include Chalfin 2010 and Billé forthcoming.

30. Chayes and Chayes 1995, 26–27.

31. Chalfin 2010, 37–51.

32. Alvarez 2009, 840–41; emphasis added. Chimni 2004 discusses international institutions constituting what he calls the "imperial global state." Koskenniemi 2001 takes a historical look at international law as the "gentle civilizer of nations." See also Pistor 2019.

33. Biersteker 1992, 130.

34. Ali 2018, 137.

35. Benton and Straumann 2010.

36. Nadelson 1992, 465; emphasis added.

37. See the full text online through the Federal Register of the United States at http://www.archives.gov/federal-register/codification/proclamations/02668.html.

38. Quoted in Nandan 1987. Delegates from Ceylon (now Sri Lanka), India, Indonesia, Japan, Kenya, and Malaysia participated in the meeting in Colombo alongside observers from Argentina, Brazil, Ecuador, and Peru.

39. The long deliberations of UNCLOS III must be seen in the context of the proposals advanced by developing countries through the New International Economic Order (NIEO). Gilman 2015 and Venzke 2017 elaborate.

40. After almost a decade of negotiations, UNCLOS III opened for signature in 1982 and came into force in 1994 upon the condition of sixty ratifications. As of this writing, there are 168 parties. While the United States recognizes the authority of UNCLOS III under international customary law, the U.S. Senate has not ratified the treaty. The Obama administration put the convention up for ratification as late as July 2012, but its passage failed.

41. See Juda 1991 for a discussion of exclusive fishery zones, which are narrowly tailored to one marine "resource" as opposed to several.

42. Nandan 1987.

43. Hunter, Salzman, and Zaelke 2007, 748.

44. Juda 1991.

45. The effort to redefine maritime sovereignty on the high seas is now under way through the new instrument called Biodiversity Beyond National Jurisdiction (BBNJ) under UNCLOS III. This treaty is poised to set new terms for economic activity beyond EEZs—half the planet's surface. I have participated in and observed some of these meetings at the United Nations headquarters in New York.

46. Nadelson 1992 discusses the legal language of "surplus."

47. For studies on the bluefin's transatlantic migrations, see Block et al. 2001.

48. Mather, Mason, and Jones 1995 detail various "stock" hypotheses.

49. Scott 1998.

50. Keohane and Nye 2001a.

51. Devnew 1983, 43–44. This section is indebted to Jack Devnew's brilliant master's thesis, which details the importance of ICCAT's early history. I thank him immensely for sharing it with me.

52. Fonteneau 2008, 10.

53. Because the Magnuson–Stevens Act had already established the fishery conservation zone of two hundred miles in March 1977, President Reagan's formal proclamation of an EEZ in March 1983 reaffirmed but did not materially alter fishery jurisdictions for the United States (Malone 1984, 803).

54. Devnew 1983, 41.

55. Captain Lou Puskas quoted in ibid., 9.

56. The National Archives have published declassified documents online, available at http://2001-2009.state.gov/documents/organization/52660.pdf. Although the Defense Department does not craft fisheries policy for the United States (the State and Commerce Departments do), UNCLOS III and claims over territorial waters still affect it.

57. See Ruais 2011–12.

58. Devnew 1983, 96.

59. August Felando quoted in ibid., 66.

60. Keohane and Nye 2001b, 135.

61. Ruais 2011–12.

62. Porch 2005, 372.

63. Ibid., 367; emphasis added.

64. Devnew 1983, 59.

65. http://www.iccat.int/Documents/BienRep/REP_BFT_82.pdf.

66. Devnew 1983, 83–85.

67. David Burney of the U.S. Tuna Foundation quoted in ibid., 74.

68. Ibid., 86.

69. http://www.iccat.int/Documents/BienRep/REP_BFT_82.pdf.

70. Bestor 2003, 304.

71. Malone 1984, 787.

72. 1980 Republican National Platform quoted in ibid., 786–87. Reagan's primary concern with UNCLOS III related to U.S. control over deep seabed minerals.

73. Aman 1995, 441.

74. Dittmer 2017, 4.

75. Alvarez 2009, 971.

76. Whynott 1995, 149.

77. Telesca 2018.

78. http://www.iccat.int/Documents/BienRep/REP_EN_96-97_I_2.pdf.

79. ICCAT Recommendation 98–5 included a provision for allocating quota to non-Contracting Parties fishing in the eastern Atlantic and Mediterranean since 1993, but did not specify them. New entrants that would receive quota later include Algeria, Albania, Egypt, Iceland, Norway, and Syria.

80. http://www.iccat.int/Documents/Recs/compendiopdf-e/1998-05-e.pdf.

81. The United Nations Fish Stock Agreement opened for signature in 1995 and entered into force in 2001 to coordinate the regulation of catch in "highly migratory fish" that "straddle" EEZs. Although the UN Fish Stocks Agreement does not directly reference historical catch, it does take into account "fishing patterns" as a "general principle" in Article 5b.

82. Singh-Renton 2010.

83. http://www.iccat.int/Documents/Recs/compendiopdf-e/2001-25-e.pdf.

84. Koskenniemi and Lehto 1996, 535.

85. Malone 1984, 803.

86. Benton, Edelstein, and the Editorial Collective 2011.

87. Weber 1964 (1947), 325.

88. Mansfield 2004; Ostrom and Schlager 1996; Porch 2005, 364.

89. Callon 1998, 3.

90. Hardin 1968, 1244–45. Hardin made some deeply problematic assumptions about personhood and behavioralism in this article but they are beyond the scope of this book to detail. An important critique of Hardin's thesis is Ostrom et al. 1999.

91. Wiber 2005, 135.

92. Hunter, Salzman, and Zaelke 2007 write: "[W]hen the exclusive economic zones were created in UNCLOS . . . it was thought that this would encourage strict national oversight to ensure conservation of domestic fish stocks. Yet exactly the opposite has occurred. Nations eager to exploit their new national

resources subsidized large fishing fleets. Between 1970 and 1990, the global fishing fleet doubled from 585,000 to 1.2 million commercial boats, in addition to the millions of small fishing boats" (763).
93. Longo and Clark 2012, 222. Urbina 2019 makes similar claims about "lawlessness" on the high seas.
94. Canfield forthcoming.
95. Longo, Clausen, and Clark 2015.
96. See the article "Managed to Death" in the October 30, 2008, issue of the *Economist* at http://www.economist.com/node/12502783?story_id=12502783. See also the blog by Andrew Revkin in the *New York Times* from November 26, 2008, titled "The (Tuna) Tragedy of the Commons."
97. Hardt and Negri 2000, 201.
98. Pistor 2019.
99. Block et al. 2001, 1313.

3. Saving the Glamour Fish

1. The phrase "charismatic megafauna" is attributed to Dennis Murphy.
2. Another fish in ICCAT's convention area that has received media attention is Atlantic swordfish. In the late 1990s, SeaWeb and the National Resources Defense Council (NRDC) launched campaigns to "Give Swordfish a Break." U.S. President Bill Clinton joined the fight and called for a ban on its sale and import in June 1998. More recently, sharks have made headlines because they too are under enormous stress. No other creature in ICCAT's convention area, however, has garnered as much sustained public attention as the bluefin.
3. The Web site for the marketplace MercaMadrid makes this claim, and dates its founding to 852 during the time of the Moors and the commercial life of the souk in Madrid. See http://www.mercamadrid.es/index.php?option=com_content&view=article&id=1028&Itemid=133.
4. My phrase "the savior plot" is inspired by the work of Lisabeth During.
5. Keck and Sikkink 1998.
6. Calhoun 2004; Hamblyn 2009.
7. Keck and Sikkink 1998.
8. Chayes and Chayes 1995, 25; emphasis in original.
9. Ibid.
10. This chapter does not presuppose the availability of a deliberative, rational discussion in which the better argument wins (Habermas 1989 [1962]) or one that is measured according to an ability to build and carry out an agenda (Gans 1979; McCombs 2005).
11. As Michel Foucault reminds us, a "regime of truth" permeates the social fabric. It is systemic. For Foucault, "truth" is not understood as or "defined by a correspondence to reality but [is] a force inherent to principles and which has to be developed in a discourse . . . [T]his truth is not something which is hidden behind or under the consciousness in the deepest and most obscure part of the soul." Truth is in plain sight. See Foucault 2007a, 163.

12. Goffman 1974, 10–15.

13. Although news outlets such as the *New York Times* need to profit from the drama they sell, the lesson is not to reduce communicative dynamics to money, advertising, and subscription. Benson and Powers 2011 and McChesney 2008 offer guidance on the political economy of news.

14. See Rosaldo 1989, 108.

15. Escobar 1993, 5. Rancière 2004 offers guidance here too.

16. Hviding 2003, 543.

17. Rosaldo 1989. The argument I present in this chapter in indebted to the probing eye of Louis Römer. I thank him immensely for helping me to see the savior plot more clearly.

18. Cowen and Smith 2009.

19. Betsill and Corell 2001; Warner 2002, 114.

20. Hardt and Negri 2000, 33.

21. The *New York Times* archives its publications for historical record keeping, which makes a search for ICCAT through the years possible, comprehensive, and reliable. To conduct my search of moments when ICCAT appeared in print or online in the *New York Times*, I cross-referenced the databases ProQuest Historical Newspapers and Lexis-Nexis through New York University–licensed access. I also searched the Web site of the *New York Times* directly. My findings are based on some stories that appeared in one database and not in others, but also on stories that appeared in two or in all three databases. Although my search was confined to one news outlet, I recognize that other media, such as documentary film and television, matter greatly. So does coverage in countries well beyond the United States. Audiences engage in personalized, interactive information gathering, which may or may not rely on mass news (Bennett 2004). The rapid rise of Twitter as a platform for environmental campaigning came after my time as a regular ICCAT delegate.

22. Powers 2018, 119–21.

23. Although the *New York Times* may influence what appears as news in other outlets, Rajagopol 2013 reminds us not to assume a U.S.-centric approach to media.

24. Parmentier 2012, 210. At the time of my field research, ICCAT charged five hundred dollars per meeting for two observers per delegation. Some environmentalists considered this an "entrance fee" and complained that the price was unreasonably high. As a delegate from the Institute for Public Knowledge (IPK) of New York University, I had to pay for each commission meeting I attended. This limited my research. The Pew Environment Group, a division of the Pew Charitable Trusts, stacked observer delegations with its own consultants, sending more representatives than the allotted two, making their participation thicker and their influence on decision makers greater relative to other environmental NGOs.

25. Powers 2018, 2–24.

26. Ibid., 156–60.

27. The comparison between IWC and ICCAT is frequent and sometimes over-

stated. The commercial importance of whales is no longer what it once was. According to a seasoned delegate with considerable experience in negotiations at both IWC and ICCAT, the moratorium on whaling means that the position for negotiation is relatively simple and straightforward: you either agree to kill the mammal or you do not. Killing fish at ICCAT is not a matter of *if*, but of *how many*.

28. Safina 1997.
29. Hamblyn 2009.
30. Whynott 1995, 152. Big thanks to Carl Safina for clarifying this history.
31. Ibid., 153.
32. Safina 1993.
33. Whynott 1995, 153–54.
34. O'Reilly 2017, 143.
35. Safina 1997, 95.
36. Another attempt to list Atlantic bluefin tuna as "endangered" erupted in 1994 when the WWF lobbied the CITES Management Authority of Kenya to initiate a proposal. Because the agency did not notify the proper authorities, Kenya withdrew it. See Ruais 2011–12.
37. Clover 2004, 26–27.
38. Safina 1993, 1997.
39. Whynott 1995, 154–57. While the National Audubon Society and the WWF led the first major bluefin tuna campaign, they did not act alone. The Center for Marine Conservation, the National Coalition for Marine Conservation, and the Council on Ocean Law also participated.
40. Safina 1993, 232–33.
41. Benton, Edelstein, and the Editorial Collective 2011.
42. Safina 1993, 233.
43. Powers 2018, 6–9.
44. Warner 2002, 94.
45. Cox 2013, 180–81.
46. See the blog by John Collins Rudolf and David Jolly from October 13, 2010, in the *New York Times*.
47. Cox 2013, 146–47.
48. In addition to Greenpeace, Oceana, the Pew Environment Group, and the WWF, other NGOs such as the Blue Ocean Institute (founded by Safina), the Center for Biological Diversity, and the Sea Shepherd Conservation Society, among others, participated in the second campaign to "save" bluefin tuna from the late 2000s.
49. See Paul Greenberg's cover story in the *New York Times Sunday Magazine*, "Tuna's End," June 27, 2010.
50. A core feature of the environmentalists' platform for regulatory action in the mid- to late 2000s was the creation of marine protected areas (MPAs), or no-fish zones protecting spawning grounds and thereby future generations of Atlantic bluefin tuna. In 2011, environmentalists dropped their demand for MPAs like a hot potato. Clarity on reasons why was not forthcoming.

51. See the report by the International Consortium of Investigative Journalists (ICIJ), "Looting the Seas: How Overfishing, Fraud & Negligence Plundered the Majestic Bluefin Tuna" (2010), at http://www.publicintegrity.org/articles/entry/2651. This report received funding from the Pew Environment Group. Reporter David Jolly documented the report's release in his blog from the *New York Times* on October 18, 2011.

52. See the blog by David Jolly in the *New York Times* dated November 27, 2010, "Group Votes to Keep Fishing Levels of Bluefin Tuna Stable."

53. See David Jolly's blog in the *New York Times* from November 29, 2010, titled "A Dirge for Bluefin Tuna."

54. Newly adopted legislation in the United States regulating gear affecting bluefin tuna prompted the editorial in the *New York Times* "Bluefin Tuna Catch a (Small) Break," April 11, 2011.

55. See the blog by David Jolly in the *New York Times* dated November 19, 2012, "Bluefin Fishing Quota Will Rise Only Slightly."

56. Ticktin 2017 offers a probing discussion of "innocence" in the parallel discourse of humanitarianism.

57. The WikiLeaks cable "09BRUSSELS1616, BARROSO'S NEW COMMISSION SHOULD BE GOOD" summarized European Commission President José Manuel Barroso's lineup of nominees for various posts in November 2009 and the extent to which they would have "good implications for U.S. economic policy interests." About Damanaki, the cable reads: "Putting a Greek in charge of anything having to do with maritime affairs . . . will certainly bring an interesting dynamic to the portfolio, and indeed to the College of Commissioners . . . [S]he is likely to support France and Spain with a continued focus on the economic problems of the EU fishing fleets rather than on conservation efforts, increasing U.S. concerns about managing endangered species and magnifying our difference on quotas for bluefin tuna and other commercially valuable catches."

58. See the *New York Times* blog by John Collins Rudolf and David Jolly, "Showdown on Bluefin Tuna," October 13, 2010.

59. See Appadurai 2013, 48.

60. Agreed in 2008, ICCAT Recommendation 08–09 was an important victory for environmentalists, who could now submit reports of noncompliance directly to the ICCAT secretariat. Environmentalists told me that this change was made possible by the leadership of Dr. William Hogarth, who once served as the ICCAT chair (and head of the U.S. delegation). Individual actors can influence ocean governance in ways big and small.

61. To my knowledge, only Greenpeace and the WWF issued paperwork about possible noncompliance to the ICCAT secretariat while I was in the field, which departs from the strategies deployed by other NGOs such as Pew at this time.

62. See the *New York Times* blog "As Regulators Meet, Fishing Boats Thumb Their Noses" by David Jolly dated June 1, 2012.

63. Mathur 2016 and Hull 2012 offer excellent guides for an anthropology of bureaucratic documents, a subject that deserves full treatment elsewhere.

64. Chayes and Chayes 1995, 25.
65. https://www.europol.europa.eu/newsroom/news/how-illegal-bluefin-tuna
-market-made-over-eur-12-million-year-selling-fish-in-spain. I imagine the
new treaty known as the Agreement on Port State Measures, which entered
into force in June 2016 to prevent and eliminate IUU fishing at port, instigated this sting operation. The EU is a signatory.
66. Warner 2002, 97.
67. According to its financial statement from 2010, based on reports by the Audit Bureau of Circulations (ABC), the *New York Times* "had the largest daily
and Sunday circulation of all seven-day newspapers in the United States . . .
In 2010, approximately . . . 74% of the Sunday circulation was sold through
subscriptions; the remainder was sold primarily on newsstands." The Sunday paper circulated to about 1,356,800 readers in 2010: https://s1.q4cdn
.com/156149269/files/doc_financials/annual/2010NYTannual.pdf.
68. Hardt and Negri 2000, 36.
69. Cronon 1996, 14.
70. DiChiro 1998, 123; Rosaldo 1989.
71. DiChiro 1998, 131.
72. Nixon 2011.
73. Ticktin 2017.
74. Plumwood 1993.

4. Alibis for Extermination

1. See the ICCAT Web site at http://www.iccat.int/en.
2. Heazle 2010.
3. Finlayson 1994, 10.
4. Jasanoff 1990, 2005; Nader 1996; Shore and Wright 1997.
5. Power 1999; Strathern 2000.
6. Hacking 2006 (1975), 1990.
7. Heazle 2010.
8. Davis et al. 2012; Merry 2011, 2016; Telesca 2017.
9. Daston 1988.
10. Ezrahi 1990.
11. Heazle 2010.
12. Hilgartner 2000.
13. Finlayson 1994, 6.
14. Hardt and Negri 2000, 31.
15. Peterson 1993, 274–75.
16. Martello and Jasanoff 2004.
17. Smith 1994 briefly discusses the origins of the term "fisheries science," which
emerged as early as 1904 in a regular column by the English scientist Joseph
Cunningham in the publication *Fish Trades Gazette*.
18. Finlayson 1994; Finley 2011; Hubbard 2006; Smith 1994.
19. Unnamed official quoted in Smith 1994, 3.
20. Ibid., 4; emphasis added.

21. Woolf 1989 discusses statistics more generally as central to the making of the modern state. For an overview of this literature from the perspective of fisheries, see Telesca 2017.
22. Anker 2002, 4–5.
23. Oreskes and Conway 2010.
24. Dr. Jake Rice quoted in Finlayson 1994, 3.
25. Ibid., 3.
26. Ellis 2008, 39.
27. For the working class, a fight with the bluefin once cost several dollars for a full or half day on a party boat. Now no longer in business, party-boat operators once consistently caught bluefin tuna just two to ten miles offshore in New Jersey. One fisher said at this January 2011 meeting in Sandy Hook that he was "effectively shut out of the fishery." U.S. regulations while I was in the field mandated that each boat dock only one bluefin tuna per day—no more. With 20, 25 or 30 anglers on board, captains could not offer each a "cube of fish and a ball of rice and [say] thanks for your three hundred bucks."
28. Merry 2016.
29. Sharp 2001.
30. https://www.iccat.int/en/index.asp; emphasis added.
31. Sibley and Ewick 2003, 79.
32. Whynott 1995, 145.
33. Sloan 2003, 77.
34. Finlayson 1994, 99.
35. Whynott 1995, 145.
36. Ibid.
37. Derman 2011; MacKenzie 2006.
38. By contrast, other tuna RFMOs such at the Inter-American Tropical Tuna Commission (IATTC) have staff that conducts "stock assessments."
39. Merry 2006, 44 expresses a similar dynamic in the context of human-rights instruments.
40. http://www.iccat.es/GBYP/en/index.htm.
41. Douglas 1966.
42. Merchant 1989 (1980), xxi. See also Plumwood 1993.
43. Dawson 2016, 52.
44. Haraway 1988. It is worth exploring more systematically elsewhere how the knowledge produced by fisheries science is linked to broader relationships of power, as the discipline of anthropology has done since the 1970s, given its ties to colonial administrations overseas, exemplified in such groundbreaking work as Asad 1973.
45. See Andrew B. Cooper, *A Guide to Fisheries Stock Assessment: From Data to Recommendations*, U.S. Department of Commerce's National Oceanic and Atmospheric Administration. I thank Steve Cadrin for sharing this material with me.
46. Edwards 2010.
47. Fisheries scientists use various models. The one adopted in the 2012 bluefin

tuna "stock assessment" was the "virtual population analysis" (VPA) model. See Porch 2005 for details.

48. Sharp 2001, 382.
49. Porch 2005, 370.
50. Porch 2005 and Safina 2001, 450–55 detail this complexity.
51. Appadurai 2016, 45.
52. Derman 2011, 156; emphasis added.
53. MacKenzie 2006.
54. Pauly 1995.
55. Kingsland 1985, 5.
56. Scott 1998, 19–22.
57. Telesca 2017; Poovey 1998, 78.
58. Poovey 2015, 223.
59. Latour 1999, 261.
60. Scott 1998, 45–46.
61. Lynch 1990; Lynch and Woolgar 1990.
62. Scott 1998, 57–58.
63. Telesca forthcoming. See Römer 2019.
64. Silbey and Ewick 2003, 77.
65. See Bauman 2000, 14.
66. Escobar 1993.

5. The Libyan Caper

1. Ortner 2006, 130. See also Ortner 1996.
2. Huizinga 1955 (1938), 13; emphasis in original.
3. Geertz 1973, 412–53.
4. Ibid., 433–36. Geertz later offered a fruitful critique of the "game" as analogy in social theory in Geertz 1980. It is summarized here in his discussion of ritual: "It can expose some of the profoundest features of social process, but at the expense of making vividly disparate matters look drably homogenous" (173).
5. Appadurai 1986, 21.
6. Geertz 1973, 433–34.
7. Moore 1978, 41.
8. Chapter 2 discusses more fully the idea of the "new sovereignty" in Chayes and Chayes 1995.
9. Bauman 2000, 123.
10. Peterson 1993, 279.
11. Buzan 1981, 325–26.
12. Wang 2010, 722.
13. See Anghie 2004.
14. http://www.greenpeace.org/international/Global/international/planet-2/report/2001/11/vote-buying-japan-s-strategy.pdf.
15. Wang 2010, 720.
16. Ostrom et al. 1999.

17. "Special" meetings of the commission happen once every two years, as do "regular" commission meetings (such as the 22d Regular Meeting of the ICCAT commission held in Istanbul, Turkey, in November 2011). The designation signifies that in "regular" ICCAT meetings matters such as the election of new chairs happen. "Special" meetings were added in 1978 so that the ICCAT commission could meet annually. What makes these meetings "special," then, is nothing at all.

18. Goffman 1972 (1955).

19. Moore 1978.

20. I learned of the proportional share of Atlantic bluefin tuna supplying the global market in a presentation by the chair of ICCAT's scientific advisory committee in 2012.

21. Issenberg 2007, 239 details the bluefin tuna trade in Libya.

22. Dittmer 2017, 128.

23. Goffman 1981.

24. Huizinga 1955 (1938), 11.

25. The chairs of ICCAT and of panels and committees greatly influenced the rulemaking process and thus what decisions ICCAT made. No wonder that among the subjects of WikiLeaks cables dedicated to ICCAT were those related to who will chair the commission. In one, Croatia indicated which candidate it supported based on who would secure its bluefin tuna quota. My observations of ICCAT meetings under the leadership of two different commission chairs demonstrated that personality and leadership style mattered considerably. According to one state delegate, "In my experience [it] can be a huge factor in how the organization operates: whether it's efficient, whether there's more or less [external] influence, whether discussions get supported in an effective way or not." There was a fine line between directing discussions and controlling them, between curbing a lengthy discussion and cutting one off prematurely, between ensuring that rules of procedure were followed and failing to accommodate the procedural needs of all delegations, such as providing time for small delegations to read draft texts before a recommendation's adoption. There was also the obvious matter of competence: Was he organized? Did he know the rules of procedure and the recommendations under review? Did he talk clearly, or did he talk too fast for the translators? Did he talk too much? Was he too assertive or too timid? Did he accurately identify the flag of a member state when it asked for the floor? To whom did he give the floor first? This balancing act bore on his status among ICCAT delegates. See the WikiLeaks cable reference number 03ZAGREB1678 titled "CROATIA DOES NOT SUPPORT U.S. CANDIDATE FOR ICCAT EXECUTIVE SECRETARY."

26. See the blog by Andrew C. Revkin titled "Group Warns of Failing Effort to Manage Tuna and Sharks" in the *New York Times* dated November 14, 2009.

27. Geertz 1973, 443.

28. Moore 1978, 64.

29. Although bluefin tuna dominated ICCAT proceedings—and media coverage—in 2010, an important item on the agenda that year was the renegotiated size

and allocation of quota for bigeye tuna, another valuable commercial fish often wrongly sold as bluefin.

30. Chinese Taipei (Taiwan) is not a Contracting Party, but retains the status of a Cooperating Non-Contracting Party, Entity or Fishing Entity (see "Cooperators" in Appendix A). This means that (1) it may be allocated a quota and (2) it must fully comply with ICCAT conservation and management measures.

31. Which countries served in the posts of chair and executive secretary carried more than symbolic importance. Elections were political decisions. According to a state delegate I interviewed in October 2012, the attitude was: "Maybe we can get someone in there so we can get more influence or information." I was not alone in observing the style of one chair as, to borrow a delegate's word, "authoritarian." He shaped the tempo of discussions, and stalled and interrupted debates in ways that could be perceived as privileging his national interests. Since ICCAT's inception, all ICCAT chairs *and* executive secretaries have been men. During its first two decades, delegates from developing countries did on occasion occupy the position of ICCAT chair. From 1991 until I was last in the field in 2012, all commission chairs were from the European Union, Japan, or the United States exclusively, with the exception of a Brazilian chair from 2007 to 2011. Leadership under Brazil was widely seen as a positive development. Countries such as Panama have since been elected to chair the commission.

32. Marcus 1983, 53.

33. Article III(7) of the ICCAT convention reads: "The official languages of the Commission shall be English, French and Spanish." In these languages, translations happened orally and in written documents, although Arabic has since been added for oral translation during commission meetings only.

34. The "precautionary principle" mandates that nation-states refrain from action that is costly to the environment should scientific proof be unavailable, noting that the onus of proof falls on those creating the threat. It has been formally adopted in several UN instruments, including the Framework Convention on Climate Change (1992) and the Rio Declaration (1992), and appears as one of the nine principles that launched the UN Global Compact (2000). It did not formally appear in the ICCAT convention at the time of my research, although there was an active debate among member states to include it in an amended convention.

35. Dittmer 2017, 14.

36. Ross 2007.

37. Huizinga 1955 (1938), 12; emphasis in original.

38. The Madrid Protocol entered into force in 2005 as a means to calculate the budgetary contributions of ICCAT member states. According to the ICCAT Web site, "This scheme divides the Contracting Parties into four groups (essentially based on classification of market economies and per capita GNP [gross national product], and on tuna catch and canned production), with every Contracting Party in each group being assigned a portion of the Com-

mission's total budget. The intent of this scheme is to reduce the financial burden on less developed countries." ICCAT's total operating budget for 2013 was €3,025,600. See http://www.iccat.int/en/finances.htm.

39. Lévi-Strauss 1976, 75–76.

40. Telesca forthcoming.

41. Neumann 2012.

42. At commission meetings, the secretariat deposited into a delegation's assigned "pigeonhole" (mailbox) several documents several times a day. An observer seeking to circulate a document had to clear it through the secretariat for distribution. The sheer volume of paperwork produced every year became so overwhelming that the commission shifted its practice in 2011. Hard copies were distributed to only three delegates from each delegation thereafter. Other delegates could reference electronic texts through a shared server. In Agadir in 2012, the anemic Internet connection at times made reviewing documents online impossible. This slowed the progress of the commission meeting enormously.

43. Mathur 2016, 4.

44. Dates and times appeared on draft documents to ensure that delegates had the most up-to-date materials for review.

45. Resolutions and recommendations did not include their sponsors when published on the ICCAT Web site, so I cannot confirm with certainty the extent to which these three trading areas have dominated authorship of ICCAT policy over the years. This matter deserves study.

46. See the article dated December 18, 2012, in the *Financial Times*, "Hollande Seeks to Smooth Algeria Ties."

47. Norway did not have an active ICCAT fishery in 2010, but it still sent four delegates to the ICCAT meeting. In other words, Norway did not have a material stake in any one ICCAT fish at the time but invested in attending ICCAT meetings anyway. ICCAT was a place where supranational regulations affecting other areas of ocean governance emerged. Norway wanted a seat at the table to ensure that its delegation contributed to discussions. Although Norway retained a small quota for bluefin tuna in the eastern Atlantic (Appendix B), it prohibited its vessels from fishing bluefin tuna in its territorial waters and on the high seas because it once considered the bluefin endangered by law. Norway wanted to retain its quota should the "stock" rebound, and to do so it had to attend ICCAT meetings. As of this writing, I understand that Norway has begun to fish its quota for bluefin tuna again.

48. The letter detailing the Norwegian objection is not part of ICCAT's official public record available online. I obtained a copy directly from the Norwegian delegation at the 2011 ICCAT meeting in Istanbul, Turkey.

49. Gambs 1946, 12–13.

50. Alvarez 2009; see also Hardt and Negri 2000 and Pistor 2019.

51. See Canfield 2018 for a discussion cautioning consensus-based, collaborative approaches to global governance as a cure-all.

Conclusion

1. In October 2018, with much fanfare, the legendary Tsukiji marketplace moved after some eighty years to a new location in Tokyo.
2. I here take inspiration from Shukin 2009, 6.
3. Farwell 2001, 399–400.
4. https://www.tokyo-zoo.net/english/kasai/main.html.
5. Malamud 1998, 234.
6. Ibid., 248.
7. Alaimo 2013.
8. Bolster 2012 offers an excellent history of overfishing in the Atlantic Ocean. For a discussion of the "shifting baseline syndrome" proposed by Daniel Pauly, see Pauly 1995.
9. https://www.washingtonpost.com/news/morning-mix/wp/2015/03/25/the-mysterious-deaths-one-by-one-of-nearly-all-the-bluefin-tuna-in-tokyo-sea-life-park/?utm_term=.f3980b51e5ad.
10. King 2013.
11. Berger 1980, 22. See Shukin 2009, 33–34 for an important critique of Berger. He risks, Shukin points out, treating the relationship between people and nonhuman animals as "primal or universal" rather than as "politically and historically contingent."
12. Berger 1980, 24; emphasis in original.
13. Williams 1977, 80.
14. Berger 1980, 28. Berger references domesticated "pets" as the primary way moderns engage with the nonhuman animal today.
15. Haraway 2008.
16. See Bauman 2000, 14.
17. Berger 1980, 26.
18. Tsing 2016, 4.
19. Telesca 2017.
20. Tsing 2016, 4.
21. Williams 1977, 82.
22. Ibid., 70–71.
23. Telesca 2018.
24. See the 2018 report by the Food and Agriculture Organization (FAO), "The State of World Fisheries and Aquaculture" available online at http://www.fao.org/state-of-fisheries-aquaculture.
25. Telesca forthcoming.
26. Williams 1977, 79–80.
27. Ibid., 83–84.
28. Berger 1980, 28.
29. ICCAT commissioned this independent performance review, available since 2009: https://www.iccat.int/Documents/Other/PERFORM_%20REV_TRI_LINGUAL.pdf.
30. Varoufakis 2017.

31. See Mitchell 2002, 53.

32. Lorde 1984.

33. Neumann 2012, 16.

34. Ibid., 124.

35. This literature is extensive. Some critiques by international legal theorists include Benvenisti and Downs 2007, Chimni 2004, and Kingsbury, Krisch, and Stewart 2005.

36. Stengers 2018, 145. When the historian Kate Brown and the microbiologist Margaret McFall-Ngai came together at an interdisciplinary conference and, like combustion, breathed oxygen into new ways of thinking about the mysterious maladies in a plutonium-manufacturing district in Russia (which Brown researches) caused not by cancer but by mutations in intestinal bacteria (which McFall-Ngai researches), we have a model for this kind of approach, as described in Tsing et al. 2017, M3–4.

37. Indigenous or traditional knowledge was never discussed at ICCAT while I was in the field. The extent to which it can be integrated into marine policy is an area worth exploring elsewhere.

38. Braithwaite 2010. The conclusion drawn from understanding pain and suffering in nonhuman natures must never be pity. Haraway 2008, 22–23 offers guidance here.

39. https://www.psychologytoday.com/blog/animal-emotions/201709/fishes-show-individual-personalities-in-response-stress.

40. Pieribone and Gruber 2005.

41. https://www.sciencedaily.com/releases/2019/02/190207142234.htm.

42. https://www.newscientist.com/article/2106331-fish-recorded-singing-dawn-chorus-on-reefs-just-like-birds/.

43. http://www.anthropocenemagazine.org/2017/06/recording-fish-song-to-make-our-fisheries-more-sustainable/.

44. Agamben 1999, 20.

45. https://www.theguardian.com/environment/2017/nov/15/plastics-found-in-stomachs-of-deepest-sea-creatures.

46. Tsing et al. 2017, G2.

47. See my blog dated July 4, 2019, at https://marenulliusblogg.w.uib.no/2019/07/04/the-language-of-ocean-governance-an-anthropological-view-of-a-global-village/.

48. https://www.weforum.org/agenda/2017/06/tuna-2020-traceability-declaration-stopping-illegal-tuna-from-coming-to-market/.

49. Urbina 2019, 8 describes the rebranded Chilean sea bass (Patagonian toothfish) as "white gold."

50. Ibid. For an excellent discussion of modern slavery, see Martínez 2011. The film *Ghost Fleet* (2018) documents the efforts to free some five thousand fishers from forced labor in the remote islands of Indonesia. In June 2014, the *Guardian* reported on the "slave labor" propping up the shrimp industry here: https://www.theguardian.com/global-development/2014/jun/10/supermarket-prawns-thailand-produced-slave-labour.

51. Farrow 2018.
52. Berger 1980, 27–28. The "thoughtlessness" of Adolf Eichmann and his heirs—the ones described by Hannah Arendt who "could not entangle, could not track the lines of living and dying, could not cultivate response-ability, could not make present to [themselves] what [they were] doing, could not live in consequences or with consequence" (Haraway 2016, 36)—is still very much in evidence with the rise of ecofascism.
53. Haraway 2016, 16.
54. Montgomery 2015. As Jacquet et al. 2019 remind us, octopus farming is an awful idea.
55. Telesca 2017.
56. Williams 1977, 82.
57. Haraway 2008, 18, 37.
58. See my blog dated November 7, 2019, at https://uminnpressblog.com/2019/11/07/cultivating-care-for-one-of-the-oceans-most-majestic-creatures/. Robin Wall Kimmerer (2013, 54–58) acknowledges the way English as a grammar limits the pronouns speakers use when referring to the sacredness of animate kin.
59. Francis 2015, 10.
60. Van Dooren 2014, 140; emphasis in original.
61. Tsing et al. 2017, G1.
62. Rosaldo 1989.
63. For models on how to affirm life in these dark times, see in anthropology Govindrajan 2018; Kirksey 2015; Raffles 2011.

Bibliography

Agamben, Giorgio. 1999. *Remnants of Auschwitz: The Witness and the Archive*. New York: Zone Books.

Agrawal, Arun. 2005. *Environmentality: Technologies of Government and the Making of Subjects*. Durham, N.C.: Duke University Press.

Alaimo, Stacy. 2013. "Violet-Black." In *Prismatic Ecology: Ecotheory beyond Green*, ed. Jeffrey Jerome Cohen, 233–51. Minneapolis: University of Minnesota Press.

Ali, Nathanael Tilahun. 2018. *Regulatory Counter-Terrorism: A Critical Appraisal of Proactive Global Governance*. New York: Routledge.

Alvarez, José Enrique. 2009. "Contemporary Foreign Investment Law: An 'Empire of Law' or the 'Law of Empire'?" *Alabama Law Review* 60: 943–74.

Aman, Alfred C., Jr. 1995. "A Global Perspective on Current Regulatory Reform: Rejection, Relocation, or Reinvention?" *Indiana Journal of Global Legal Studies* 2: 429–64.

Anderson, Benedict. 2002 (1983). *Imagined Communities: Reflections on the Origin and Spread of Nationalism*. New York: Verso.

Anghie, Antony. 2004. *Imperialism, Sovereignty and the Making of International Law*. New York: Cambridge University Press.

Anker, Peder. 2002. *Imperial Ecology: Environmental Order in the British Empire, 1895–1945*. Cambridge, Mass.: Harvard University Press.

Appadurai, Arjun. 1986. "Introduction: Commodities and the Politics of Value." In *The Social Life of Things: Commodities in Cultural Perspective*, ed. Arjun Appadurai, 3–63. New York: Cambridge University Press.

——. 1996. *Modernity at Large: Cultural Dimensions of Globalization*. Minneapolis: University of Minnesota Press.

——. 2013. *The Future as Cultural Fact: Essays on the Global Condition*. New York: Verso.

——. 2016. *Banking on Words: The Failure of Language in the Age of Derivative Finance*. Chicago: University of Chicago Press.

Asad, Talal, ed. 1973. *Anthropology and the Colonial Encounter*. Atlantic Highlands, N.J.: Humanities Press.

Bailey, Kevin M. 2013. *Billion-Dollar Fish: The Untold Story of Alaska Pollock*. Chicago: University of Chicago Press.

Ballestero, Andrea. 2019. *A Future History of Water*. Durham, N.C.: Duke University Press.

Baptist, Edward E. 2016. "Toward a Political Economy of Slave Labor: Hands, Whipping-Machines, and Modern Power." In *Slavery's Capitalism: A New History of American Economic Development*, ed. Sven Beckert and Seth Rockman, 31–61. Philadelphia: University of Pennsylvania Press.

Barnett, Michael, and Martha Finnemore. 2004. *Rules for the World: International Organizations in Global Politics*. Ithaca, N.Y.: Cornell University Press.

Bauman, Zygmunt. 2000. *Liquid Modernity*. Malden, Mass.: Polity Press.

Bavington, Dean. 2010. *Managed Annihilation: An Unnatural History of the Newfoundland Cod Collapse*. Vancouver: University of British Columbia Press.

Bear, Laura, and Nayanika Mathur. 2015. "Remaking the Public Good: A New Anthropology of Bureaucracy." *Cambridge Journal of Anthropology* 33(1): 18–34.

Beckert, Sven. 2014. *Empire of Cotton: A Global History*. New York: Vintage Books.

Bekker-Nielsen, Tonnes. 2005. "Introduction." In *Ancient Fishing and Fish Processing in the Black Sea Region*, ed. Tonnes Bekker-Nielsen, 13–19. Oakville, Conn.: Aarhus University Press.

Bekoff, Marc. 2004. "Wild Justice and Fair Play: Cooperation, Forgiveness, and Morality in Animals." *Biology and Philosophy* 19: 489–520.

Bennett, W. Lance. 2004. "Global Media and Politics: Transnational Communication Regimes and Civic Cultures." *Annual Review of Political Science* 7: 125–48.

Benson, Rodney, and Matthew Powers. 2011. "Public Media and Political Independence." *Public Policy Report*. Washington, D.C.: Free Press.

Benton, Lauren, and Benjamin Straumann. 2010. "Acquiring Empire by Law: From Roman Doctrine to Early Modern European Practice." *Law and History Review* 28: 1–38.

Benton, Lauren, Dan Edelstein, and the Editorial Collective. 2011. "Of Pirates, Empire, and Terror: An Interview with Lauren Benton and Dan Edelstein." *Humanity* 2: 75–84.

Benvenisti, Eyal, and George W. Downs. 2007. "The Empire's New Clothes: Political Economy and the Fragmentation of International Law." *Stanford Law Review* 60(2): 595–631.

Berger, John. 1980. *About Looking*. New York: Vintage International.

Bestor, Theodore C. 2000. "How Sushi Went Global." *Foreign Policy* 121: 54–63.

——. 2003. "Markets and Places: Tokyo and the Global Tuna Trade." In *The Anthropology of Space and Place: Locating Culture*, ed. Setha M. Low and Denise Lawrence-Zúñiga, 301–20. Malden, Mass.: Blackwell.

——. 2004. *Tsukiji: The Fish Market at the Center of the World*. Berkeley: University of California Press.

Betsill, Michele M., and Elizabeth Corell. 2001. "NGO Influence in International Environmental Negotiations: A Framework for Analysis." *Global Environmental Politics* 1: 65–85.

Biersteker, Thomas J. 1992. "The 'Triumph' of Neoclassical Economics in the Developing World: Policy Convergence and Bases of Governance in the International Economic Order." In *Governance without Government: Order and Change in World Politics*, ed. James N. Rosenau and Ernst-Otto Czempiel, 102–31. New York: Cambridge University Press.

Billé, Franck. 2016. "Introduction to 'Cartographic Anxieties.'" *Cross-Currents: East Asian History and Culture Review* 21: 1–18.

———, ed. Forthcoming. *Voluminous States: Sovereignty, Materiality, and the Territorial Imagination*. Durham, N.C.: Duke University Press.

Block, Barbara A., Heidi Dewar, Susanna B. Blackwell, Thomas D. Williams, Eric D. Prince, Charles J. Farwell, Andre Boustany, Steven L. H. Teo, Andrew Seitz, Andreas Walli, and Douglas Fudge. 2001. "Migratory Movements, Depth Preferences, and Thermal Biology of Atlantic Bluefin Tuna." *Science* 293(5533): 1310–14.

Block, Fred L. 2001. "Introduction." In *The Great Transformation: The Political and Economic Origins of Our Time*, by Karl Polanyi, xviii–xxxviii. Boston: Beacon Press.

Bolster, W. Jeffrey. 2012. *The Mortal Sea: Fishing the Atlantic in the Age of Sail*. Cambridge, Mass.: Harvard University Press.

Boltanski, Luc, and Eve Chiapello. 2005. *The New Spirit of Capitalism*. Trans. Gregory Elliot. New York: Verso.

Bonanno, Alessandro, and Douglas Constance. 1996. *Caught in the Net: The Global Tuna Industry, Environmentalism, and the State*. Lawrence: University Press of Kansas.

Bonneuil, Christophe, and Jean-Baptiste Fressoz. 2017. *The Shock of the Anthropocene*. New York: Verso.

Borch, Christian, and Robert Wosnitzer, eds. Forthcoming. *Handbook of Critical Finance Studies*. New York: Routledge.

Borgström, Georg. 1964. *Japan's World Success in Fishing*. London: Fishing News.

Braithwaite, Victoria. 2010. *Do Fish Feel Pain?* New York: Oxford University Press.

Braund, David. 1995. "Fish from the Black Sea: Classical Byzantium and the Greekness of Trade." In *Food in Antiquity*, ed. John Wilkins, David Harvey, and Michael Dobson, 162–71. Exeter: University of Exeter Press.

Brown, John Pairman. 1968. "Cosmological Myth and the Tuna of Gibraltar." *Transactions and Proceedings of the American Philological Association* 99: 37–62.

Buzan, Barry. 1981. "Negotiating by Consensus: Developments in Technique at the United Nations Conference on the Law of the Sea." *American Journal of International Law* 75: 324–48.

Calhoun, Craig. 2004. "A World of Emergencies: Fear, Intervention, and the Limits of Cosmopolitan Order: The 35th Sorokin Lecture." *Canadian Review of Sociology and Anthropology* 41: 373–95.

———. 2007. *Nations Matter: Culture, History and the Cosmopolitan Dream*. New York: Routledge.

Callon, Michel. 1998. "Introduction: The Embeddedness of Economic Markets in Economics." In *The Laws of the Markets*, ed. Michel Callon, 1–57. Malden, Mass.: Blackwell.

Canfield, Matthew C. 2018. "Disputing the Global Land Grab: Claiming Rights and Making Markets through Collaborative Governance." *Law and Society Review* 52(4): 994–1025.

———. Forthcoming. "Property Regimes." In *Oxford Handbook on Law and Anthropology*, ed. Mark Goodale, Marie-Claire Foblets, Maria Sapignoli, and Olaf Zenker. New York: Oxford University Press.

Carruthers, Bruce G., and Wendy Nelson Espeland. 1991. "Accounting for Rationality: Double-Entry Bookkeeping and the Rhetoric of Economic Rationality." *American Journal of Sociology* 97(1): 31–69.

Carson, Rachel L. 1989 (1950). *The Sea Around Us*. New York: Oxford University Press.

Cassady, Ralph. 1974. *Exchange by Private Treaty*. Austin: University of Texas Press.

Cassese, Sabino. 2005. "Administrative Law without the State? The Challenge of Global Regulation." *New York University Journal of International Law and Politics* 37: 663–94.

Ceballos, Gerardo, and Paul R. Ehrlich. 2018. "The Misunderstood Sixth Mass Extinction." *Science* 360(6393): 1080–81.

Ceballos, Gerardo, Paul R. Ehrlich, Anthony D. Barnosky, Andrés García, Robert M. Pringle, and Todd M. Palmer. 2015. "Accelerated Modern Human-Induced Species Losses: Entering the Sixth Mass Extinction." *Science Advances* 1(5): 1–5.

Chakrabarty, Dipesh. 2009. "The Climate of History: Four Theses." *Critical Inquiry* 35: 197–222.

———. 2017. "The Politics of Climate Change Is More Than the Politics of Capitalism." *Theory, Culture and Society* 34(2–3): 25–37.

Chalfin, Brenda. 2010. *Neoliberal Frontiers: An Ethnography of Sovereignty in West Africa*. Chicago: University of Chicago Press.

Chayes, Abram, and Antonia Handler Chayes. 1995. *The New Sovereignty: Compliance with International Regulatory Agreements*. Cambridge, Mass.: Harvard University Press.

Chimni, B. S. 2004. "International Institutions Today: An Imperial Global State in the Making." *European Journal of International Law* 15(1): 1–37.

Clarke, Adele E., and Donna Haraway, eds. 2018. *Making Kin Not Population*. Chicago: Prickly Paradigm Press.

Clover, Charles. 2004. *The End of the Line: How Overfishing is Changing the World and What We Eat*. London: Ebury Press.

Collette, Bruce B., Carol Reeb, and Barbara A. Block. 2001. "Systematics of the Tunas and Mackerels (Scombridae)." In *Tuna: Physiology, Ecology, and Evolution*, ed. Barbara A. Block and E. Donald Stevens, 1–33. New York: Academic Press.

Comaroff, Jean, and John L. Comaroff. 2005. "Beasts, Banknotes and the Color of Money in Colonial South Africa." *Archeological Dialogues* 12(2): 107–32.

Connery, Christopher L. 2001. "Ideologies of Land and Sea: Alfred Thayer Mahan, Carl Schmitt and the Shaping of Global Myth Elements." *boundary* 2(28): 173–201.

Cooper, Fred. 2005. *Colonialism in Question: Theory, Knowledge, History.* Berkeley: University of California Press.

Corson, Trevor. 2007. *The Zen of Fish: The Story of Sushi, from Samurai to Supermarket.* New York: HarperCollins.

Cowen, Deborah, and Neil Smith. 2009. "After Geopolitics? From the Geopolitical Social to Geoeconomics." *Antipode* 41: 22–48.

Cox, J. Robert. 2013. *Environmental Communication and the Public Sphere.* 3d ed. Washington, D.C.: Sage.

Cronon, William. 1996. "The Trouble with Wilderness; or, Getting Back to the Wrong Nature." *Environmental History* 1(1): 7–28.

Cullis-Suzuki, Sarika, and Daniel Pauly. 2010. "Failing the High Seas: A Global Evaluation of Regional Fisheries Management Organizations." *Marine Policy* 34: 1036–42.

Curtis, R. 1988. "Spanish Trade in Salted Fish Products in the First and Second Centuries AD." *International Journal of Nautical Archeology and Underwater Exploration* 17: 205–10.

Cushing, David H. 1988. *The Provident Sea.* New York: Cambridge University Press.

Daniel, Hawthorne, and Francis Minot. 1961 (1954). *The Inexhaustible Sea.* New York: Collier Books.

Daston, Lorraine. 1988. *Classical Probability in the Enlightenment.* Princeton, N.J.: Princeton University Press.

Davis, Kevin, Angelina Fisher, Benedict Kingsbury, and Sally Engle Merry. 2012. *Governance by Indicators: Global Power through Quantification and Rankings.* Oxford: Oxford University Press.

Dawson, Ashley. 2016. *Extinction: A Radical History.* New York: OR Books.

de la Cadena, Marisol, and Mario Blaser, eds. 2018. *A World of Many Worlds.* Durham, N.C.: Duke University Press.

Deloughrey, Elizabeth. 2019. "Toward a Critical Ocean Studies for the Anthropocene." *English Language Notes* 57(1): 21–36.

Demos, T. J. 2017. *Against the Anthropocene: Visual Culture and Environment Today.* Berlin: Sternberg Press.

Derman, Emanuel. 2011. *Models. Behaving. Badly. Why Confusing Illusion with Reality Can Lead to Disaster, on Wall Street and in Life.* New York: Free Press.

DeSombre, Elizabeth R., and J. Samuel Barkin. 2011. *Fish.* Malden, Mass.: Polity.

de Sousa Santos, Boaventura. 1987. "Law: A Map of Misreading. Toward a Postmodern Conception of Law." *Journal of Law and Society* 14(3): 279–302.

Devnew, Jack. 1983. "Politics and the Conservation of Atlantic Bluefin Tuna." MA thesis, Department of Marine Policy, University of Delaware.

Dezalay, Yves, and Bryant G. Garth. 1995. *The Internationalization of Palace Wars: Lawyers, Economists, and the Contest to Transform Latin American States.* Chicago: University of Chicago Press.

DiChiro, Giovanna. 1998. "Nature as Community: The Convergence of Environment and Social Justice." In *Privatizing Nature: Political Struggles for the Global*

Commons, ed. Michael Goldman, 120–43. New Brunswick, N.J.: Rutgers University Press.

———. 2016. "Environmental Justice and the Anthropocene Meme." In *The Oxford Handbook of Environmental Political Theory*, ed. Teena Gabrielson, Cheryl Hall, John M. Meyer, and David Schlosberg, 362–81. Oxford: Oxford University Press.

Di Natale, Antonio. 2012. "An Iconography of Tuna Traps. Essential Information for the Understanding of the Technological Evolution of This Ancient Fishery." *Collect. Vol. Sci. Pap. ICCAT* 67(1): 33–74.

Dittmer, Jason. 2017. *Diplomatic Material: Affect, Assemblage, and Foreign Policy.* Durham, N.C.: Duke University Press.

Douglas, Mary. 1966. *Purity and Danger: An Analysis of Concepts of Pollution and Taboo.* New York: Frederick A. Praeger.

Edwards, Paul N. 2010. *A Vast Machine: Computer Models, Climate Data and the Politics of Global Warming.* Cambridge, Mass.: MIT Press.

Ellis, Richard. 2008. *Tuna: A Love Story.* New York: Alfred A. Knopf.

Endo, Aiko, and Masahiro Yamao. 2007. "Policies Governing the Distribution of By-Products from Scientific and Small-Scale Coastal Whaling in Japan." *Marine Policy* 31: 169–81.

Epstein, Charlotte. 2008. *The Power of Words in International Relations: Birth of an Anti-Whaling Discourse.* Cambridge, Mass.: MIT Press.

Escobar, Arturo. 1993. *Encountering Development: The Making and Unmaking of the Third World.* Princeton, N.J.: Princeton University Press.

Ezrahi, Yaron. 1990. *The Descent of Icarus: Science and the Transformation of Contemporary Democracy.* Cambridge, Mass.: Harvard University Press.

Fagan, Brian M. 2006. *Fish on Friday: Feasting, Fasting, and the Discovery of the New World.* New York: Basic Books.

Farrow, Ronan. 2018. *War on Peace: The End of Diplomacy and the Decline of American Influence.* New York: W. W. Norton.

Farwell, Charles J. 2001. "Tunas in Captivity." In *Tuna: Physiology, Ecology, and Evolution*, ed. Barbara A. Block and E. Donald Stevens, 391–413. New York: Academic Press.

Feldman, Gregory. 2011. "Illuminating the Apparatus: Steps toward a Nonlocal Ethnography of Global Governance." In *Policy Worlds: Anthropology and the Analysis of Contemporary Power*, ed. Cris Shore, Susan Wright, and Davide Però, 32–49. New York: Berghahn Books.

Ferguson, James. 1988. "Cultural Exchange: New Developments in the Anthropology of Commodities." *Cultural Anthropology* 3: 488–513.

———. 1990. *The Anti-Politics Machine: "Development," Depoliticization and Bureaucratic Power in Lesotho.* New York: Cambridge University Press.

Finlayson, Alan Christopher. 1994. *Fishing for Truth: A Sociological Analysis of Northern Cod Stock Assessments from 1977 to 1990.* St. John's, Newfoundland: Institute of Social and Economic Research.

Finley, Carmel. 2011. *All the Fish in the Sea: Maximum Sustainable Yield and the Failure of Fisheries Management*. Chicago: University of Chicago Press.

———. 2017. *All the Boats in the Ocean: How Government Subsidies Led to Global Overfishing*. Chicago: University of Chicago Press.

Fonteneau, Alain. 2008. "Scientific and Historical Summary of ICCAT." ICCAT 40th Anniversary 1966–2006. Madrid: ICCAT.

Foucault, Michel. 2007a. *The Politics of Truth*. Los Angeles: Semiotext(e).

———. 2007b. *Security Territory, Population: Lectures at the Collège of France, 1977–78*. Trans. G. Burchell. New York: Palgrave Macmillan.

———. 2008. *The Birth of Biopolitics: Lectures at the Collège of France, 1978–79*. Trans. G. Burchell. New York: Palgrave Macmillan.

Francis. 2015. *Encyclical on Climate Change and Inequality: On Care for Our Common Home*. Introduction by Naomi Oreskes. Brooklyn: Melville House.

Franklin, Sarah. 2006. "Bio-economies: Biowealth from the Inside Out." *Development* 49(4): 97–101.

French, Lindsay. 1999. "Hierarchies of Value at Angkor Wat." *Ethnos* 64(2): 170–91.

Fromentin, Jean-Marc, and Joseph E. Powers. 2005. "Atlantic Bluefin Tuna: Population Dynamics, Ecology, Fisheries and Management." *Fish and Fisheries* 6: 281–306.

Gambs, John S. 1946. *Beyond Supply and Demand: A Reappraisal of Institutional Economics*. New York: Columbia University Press.

Gans, Herbert J. 1979. *Deciding What's News*. New York: Pantheon.

Geertz, Clifford. 1973. *The Interpretation of Cultures*. New York: Basic Books.

———. 1978. "The Bazaar Economy: Information and Search in Peasant Marketing." *American Economic Review* 68: 28–32.

———. 1980. "Blurred Genres: The Refiguration of Social Thought." *American Scholar* 49(2): 165–79.

———. 1983. *Local Knowledge: Further Essays in Interpretive Anthropology*. New York: Basic Books.

Ghosh, Amitav. 2016. *The Great Derangement*. Chicago: University of Chicago Press.

Gilman, Nils. 2015. "The New International Economic Order: A Reintroduction." *Humanity: An International Journal of Human Rights, Humanitarianism, and Development* 6: 1–16.

Gilroy, Paul. 1993. *The Black Atlantic: Modernity and Double Consciousness*. Cambridge, Mass.: Harvard University Press.

Goffman, Erving. 1972 (1955). "On Face Work: An Analysis of Ritual Elements in Social Interaction." In *Communication in Face to Face Interaction: Selected Readings*, ed. J. Laver and S. Hutcheson, 319–46. Harmondsworth: Penguin.

———. 1974. *Frame Analysis: An Essay on the Organization of Experience*. New York: Harper Colophon Books.

———. 1981. *Forms of Talk*. Philadelphia: University of Pennsylvania Press.

Gómez-Barris, Macarena. 2017. *The Extractive Zone: Social Ecologies and Decolonial Perspectives*. Durham, N.C.: Duke University Press.

Gordon, Colin. 1991. "Governmental Rationality: An Introduction." In *The Foucault Effect: Studies in Governmentality*, ed. Graham Burchell, Colin Gordon, and Peter Miller, 1–52. Chicago: University of Chicago Press.

Govindrajan, Radhika. 2018. *Animal Intimacies: Interspecies Relatedness in India's Central Himalayas*. Chicago: University of Chicago Press.

Graeber, David. 2001. *Toward an Anthropological Theory of Value: The False Coin of Our Own Dreams*. New York: Palgrave.

Graham, Jeffrey B., and Kathryn A. Dickson. 2001. "Anatomical and Physiological Specializations for Ethnothermy." In *Tuna: Physiology, Ecology, and Evolution*, ed. Barbara A. Block and E. Donald Stevens, 121–65. New York: Academic Press.

Grandin, Greg. 2014. *The Empire of Necessity: Slavery, Freedom, and Deception in the New World*. New York: Metropolitan Books.

Greenberg, Paul. 2010. *Four Fish: The Future of the Last Wild Food*. New York: Penguin Books.

Grusin, Richard, ed. 2017. *Anthropocene Feminism*. Minneapolis: University of Minnesota Press.

——, ed. 2018. *After Extinction*. Minneapolis: University of Minnesota Press.

Habermas, Jürgen. 1989 (1962). *The Structural Transformation of the Public Sphere: An Inquiry into a Category of Bourgeois Society*. Trans. Thomas Burger and Frederick Lawrence. Cambridge, Mass.: MIT Press.

Hacking, Ian. 1990. *The Taming of Chance*. New York: Cambridge University Press.

——. 2006 (1975). *The Emergence of Probability: A Philosophical Study of Early Ideas about Probability, Induction and Statistical Inference*. 2d ed. New York: Cambridge University Press.

Halliday, Terence C., and Bruce G. Carruthers. 2009. *Bankrupt: Global Lawmaking and Systemic Financial Crisis*. Palo Alto, Calif.: Stanford University Press.

Hamblyn, Richard. 2009. "The Whistleblower and the Canary: Rhetorical Constructions of Climate Change." *Journal of Historical Geography* 35: 223–36.

Haraway, Donna J. 1988. "Situated Knowledges: The Science Question in Feminism and the Privilege of Partial Perspective." *Feminist Studies* 14: 575–99.

——. 2008. *When Species Meet*. Minneapolis: University of Minnesota Press.

——. 2016. *Staying with the Trouble: Making Kin in the Chthulucene*. Durham, N.C.: Duke University Press.

Hardin, Garrett. 1968. "The Tragedy of the Commons." *Science* 162: 1243–48.

Hardt, Michael, and Antonio Negri. 2000. *Empire*. Cambridge, Mass.: Harvard University Press.

Harrington, Jennie M., Ransom A. Myers, and Andrew A. Rosenberg. 2005. "Wasted Fishery Resources: Discarded By-Catch in the USA." *Fish and Fisheries* 6(4): 350–61.

Harvey, David. 2006 (1982). *The Limits to Capital*. New York: Verso.

Hawkin, Paul, Amory Lovens, and L. Hunter Lovens. 1999. *Natural Capitalism: Creating the Next Industrial Revolution.* Boston: Little, Brown.

Heazle, Michael. 2010. *Uncertainty in Policy Making: Values and Evidence in Complex Decisions.* New York: Routledge.

Heise, Ursula K. 2016. *Imagining Extinction: The Cultural Meanings of Endangered Species.* Chicago: University of Chicago Press.

Helmreich, Stefan. 2008. "Species of Biocapital." *Science as Culture* 17: 463–78.

Hilgartner, Stephen. 2000. *Science on Stage: Expert Advice as Public Drama.* Palo Alto, Calif.: Stanford University Press.

Hitlin, Steven, and Jane Allyn Piliavin. 2004. "Values: Reviving a Dormant Concept." *Annual Review of Sociology* 30: 359–93.

Holling, C. S. 1973. "Resilience and Stability of Ecological Systems." *Annual Review of Ecology and Systematics* 4: 1–23.

Holm, Poul. 2012. "World War II and the 'Great Acceleration' of North Atlantic Fisheries." *Global Environment* 10: 66–91.

Holms, Douglas R., and George E. Marcus. 2005. "Cultures of Expertise and the Management of Globalization." In *Global Assemblages: Technology, Politics, and Ethics as Anthropological Problems,* ed. Aihwa Ong and Stephan J. Collier, 235–52. Malden, Mass.: Blackwell Publishing.

Holt, Sidney. 2006. "The Notion of Sustainability." In *Gaining Ground: In Pursuit of Ecological Sustainability,* ed. David M. Lavigne, 43–82. London: International Fund for Animal Welfare.

Horden, Peregrine, and Nicholas Purcell. 2000. *The Corrupting Sea: A Study of Mediterranean History.* Malden, Mass.: Blackwell Publishers.

Hubbard, Jennifer M. 2006. *A Science on the Scales: The Rise of Canadian Atlantic Fisheries Biology, 1898–1939.* Toronto: University of Toronto Press.

Hughes, David McDermott. 2017. *Energy without Conscience: Oil, Climate Change, and Complicity.* Durham, N.C.: Duke University Press.

Huizinga, Johan. 1955 (1938). *Homo Ludens: A Study of the Play-Element in Culture.* Boston: Beacon Press.

Hull, Matthew. S. 2012. "Documents and Bureaucracy." *Annual Review of Anthropology* 41: 251–67.

Hunter, David, James Salzman, and Durwood Zaelke. 2007. *International Environmental Law and Policy.* New York: Foundation Press.

Hviding, Edvard. 2003. "Contested Rainforests, NGOs, and Projects of Desire in Solomon Islands." *International Social Science Journal* 55(178): 539–54.

Ingold, Tim. 2012. "Toward an Ecology of Materials." *Annual Review of Anthropology* 41: 427–42.

Innis, Harold A. 1940. *The Cod Fisheries: The History of an International Economy.* New Haven, Conn.: Yale University Press.

International Commission for the Conservation of Atlantic Tunas. 2009. "Report of the Independent Performance Review of ICCAT." Madrid: ICCAT. https://www.iccat.int/Documents/Other/PERFORM_%20REV_TRI_LINGUAL.pdf.

Issenberg, Sasha. 2007. *The Sushi Economy: Globalization and the Making of a Modern Delicacy.* New York: Gotham Books.

Jacquet, Jennifer, Becca Franks, Peter Godfrey-Smith, and Walter Sánchez-Suárez. 2019. "The Case against Octopus Farming." *Issues in Science and Technology* 35(2): 37–44.

Jasanoff, Sheila. 1990. *The Fifth Branch: Science Advisers as Policymakers.* Cambridge, Mass.: Harvard University Press.

———. 2005. *Designs on Nature: Science and Democracy in Europe and the United States.* Princeton, N.J.: Princeton University Press.

Johnston, Bruce F. 1953. *Japanese Food Management in World War II.* Stanford, Calif.: Stanford University Press.

Joseph, May. 2019. *Sea Log: Indian Ocean to New York.* New York: Routledge.

Juda, Lawrence. 1991. "World Marine Fish Catch in the Age of Exclusive Economic Zones and Exclusive Fishery Zones." *Ocean Development and International Law* 22: 1–32.

Keck, Margaret E., and Kathryn Sikkink. 1998. *Activists beyond Borders: Advocacy Networks in International Politics.* Ithaca, N.Y.: Cornell University Press.

Keohane, Robert O., and Joseph S. Nye. 1974. "Transgovernmental Relations and International Organizations." *World Politics* 27: 39–62.

———. 2001a. "Between Centralization and Fragmentation: The Club Model of Multilateral Cooperation and the Problems of Democratic Legitimacy." KSG Working Paper No. 01-004. Available at SSRN: http://ssrn.com/abstract=262175 or http://dx.doi.org/10.2139/ssrn.26217.

———. 2001b. *Power and Interdependence: Third Edition.* New York: Longman.

Kimmerer, Robin Wall. 2013. *Braiding Sweetgrass: Indigenous Wisdom, Scientific Knowledge, and the Teachings of Plants.* Minneapolis: Milkweed Editions.

King, Barbara. 2013. *How Animals Grieve.* Chicago: University of Chicago Press.

Kingsbury, Benedict. 1998. "The Concept of Compliance as a Function of Competing Conceptions of International Law." *Michigan Journal of International Law* 19: 345–72.

Kingsbury, Benedict, Nico Krisch, and Richard B. Stewart. 2005. "The Emergence of Global Administrative Law." *Law and Contemporary Problems* 68: 15–61.

Kingsland, Sharon E. 1985. *Modeling Nature: Episodes in the History of Population Ecology.* Chicago: University of Chicago Press.

Kirksey, Eben. 2015. *Emergent Ecologies.* Durham, N.C.: Duke University Press.

Kirksey, S. Eben, and Stefan Helmreich. 2010. "The Emergence of Multispecies Ethnography." *Cultural Anthropology* 25: 545–76.

Knecht, G. Bruce. 2006. *Hooked: Pirates, Poaching, and the Perfect Fish.* Emmaus, Pa.: Rodale.

Kolbert, Elizabeth. 2014. *The Sixth Extinction: An Unnatural History.* New York: Henry Holt.

Kopytoff, Igor. 1986. "The Cultural Biography of Things: Commoditization as Process." In *The Social Life of Things: Commodities in Cultural Perspective,* ed. Arjun Appadurai, 64–91. New York: Cambridge University Press.

Koskenniemi, Martti. 2001. *The Gentle Civilizer of Nations: The Rise and Fall of International Law, 1870–1960.* New York: Cambridge University Press.

———. 2006. *From Apology to Utopia: The Structure of International Legal Argument.* New York: Cambridge University Press.

Koskenniemi, Martti, and Marja Lehto. 1996. "The Privilege of Universality: International Law, Economical Ideology and Seabed Resources." *Nordic Journal of International Law* 65: 533–55.

Kurlansky, Mark. 1998. *Cod: A Biography of the Fish That Changed the World.* New York: Penguin Books.

Larkin, P. A. 1977. "An Epitaph for the Concept of Maximum Sustainable Yield." *Transactions of the American Fisheries Society* 106(1): 1–11.

Latour, Bruno. 1999. *Pandora's Hope: Essays on the Reality of Science Studies.* Cambridge, Mass.: Harvard University Press.

Levinson, Marc. 2006. *The Box: How the Shipping Container Made the World Smaller and the World Economy Bigger.* Princeton, N.J.: Princeton University Press.

Lévi-Strauss, Claude. 1976. "The Science of the Concrete." In *Ritual, Play and Performance: Readings in the Social Sciences/Theatre,* ed. Richard Schechner and Mady Schuman, 74–76. New York: Seabury Press.

Lewis, Simon L., and Mark A. Maslin. 2015. "A Transparent Framework for Defining the Anthropocene Epoch." *Anthropocene Review* 2(2): 128–46.

LiPuma, Edward, and Benjamin Lee. 2004. *Financial Derivatives and the Globalization of Risk.* Durham, N.C.: Duke University Press.

Longo, Stefano B., and Brett Clark. 2012. "The Commodification of Bluefin Tuna: The Historical Transformation of the Mediterranean Fishery." *Journal of Agrarian Change* 12: 204–26.

Longo, Stefano B., Rebecca Clausen, and Brett Clark. 2015. *The Tragedy of the Commodity: Oceans, Fisheries, and Aquaculture.* New Brunswick, N.J.: Rutgers University Press.

Lorde, Audre. 1984. *Sister Outsider.* Berkeley: Ten Speed Press.

Lorimer, Jaime. 2015. *Wildlife in the Anthropocene: Conservation after Nature.* Minneapolis: University of Minnesota Press.

Luttwak, Edward N. 1990. "From Geopolitics to Geo-Economics: Logic of Conflict, Grammar of Commerce." *National Interest* 20: 17–23.

Lynch, Michael. 1990. "The Externalized Retina: Selection and Mathematization in the Visual Documentation of Objects in the Life Sciences." In *Representation in Scientific Practice,* ed. Michael Lynch and Steve Woolgar, 153–86. Cambridge, Mass.: MIT Press.

Lynch, Michael, and Steve Woolgar. 1990. "Introduction: Sociological Orientations to Representational Practice in Science." In *Representation in Scientific Practice,* ed. Michael Lynch and Steve Woolgar, 1–18. Cambridge, Mass.: MIT Press.

Lytle, E. 2006. "Marine Fisheries and the Ancient Greek Economy." PhD thesis, Department of Classical Studies, Duke University.

MacKenzie, Donald. 2006. *An Engine, Not a Camera: How Financial Models Shape Markets.* Cambridge, Mass.: MIT Press.

Maggio, Theresa. 2000. *Mattanza: The Ancient Sicilian Ritual of Bluefin Tuna Fishing.* New York: Penguin Books.

Malamud, Randy. 1998. *Reading Zoos: Representations of Animals and Captivity.* New York: New York University Press.

Malm, Andreas, and Alf Hornborg. 2014. "The Geology of Mankind? A Critique of the Anthropocene Narrative." *Anthropocene Review* 1(1): 62–69.

Malone, James L. 1984. "The United States and the Law of the Sea." *Virginia Journal of International Law* 24: 785–807.

Mancke, Elizabeth. 1999. "Early Modern Expansion and the Politicization of Oceanic Space." *Geographical Review* 89: 225–36.

Mansfield, Becky. 2004. "Neoliberalism in the Oceans: 'Rationalization,' Property Rights and the Commons Question." *Geoforum* 35: 313–26.

Marcus, George E. 1983. "Elite Communities and Institutional Orders." In *Elites: Ethnographic Issues,* ed. George Marcus, 41–58. Albuquerque: University of New Mexico Press.

———. 1995. "Ethnography in/of the World System: The Emergence of Multi-Sited Ethnography." *Annual Review of Anthropology* 24: 95–117.

Marks, Susan. 2003. "Empire's Law." *Indiana Journal of Global Legal Studies* 10: 449–66.

———. 2009. "False Contingency." *Current Legal Problems* 62(1): 1–21.

Martello, Marybeth Long, and Sheila Jasanoff. 2004. "Introduction: Globalization and Environmental Governance." In *Earthly Politics: Local and Global in Environmental Governance,* ed. Sheila Jasanoff and Marybeth Long Martello, 1–29. Cambridge, Mass.: MIT Press.

Martin, Randy. 2009. "The Twin Towers of Financialization: Entanglements of Political and Cultural Economies." *Global South* 3: 108–25.

Martínez, Samuel. 2011 "'Taking Better Account': Contemporary Slavery, Gendered Narratives and the Feminization of Struggle." *Humanity* 2(2): 277–303.

Massumi, Brian. 2018. *99 Theses on the Revaluation of Value: A Postcapitalist Manifesto.* Minneapolis: University of Minnesota Press.

Mather, Frank J., John M. Mason Jr., and Albert C. Jones. 1995. *Historical Document: Life History and Fisheries of Atlantic Bluefin Tuna.* Miami: U.S. Department of Commerce, National Oceanographic and Atmospheric Administration (NOAA) Technical Memorandum, NMFS-SEFSC 370.

Mathur, Nayanika. 2016. *Paper Tiger: Law, Bureaucracy and the Developmental State in Himalayan India.* New York: Cambridge University Press.

McChesney, Robert. 2008. *The Political Economy of Media.* New York: Monthly Review Press.

McCombs, Maxwell. 2005. "A Look at Agenda-Setting: Past, Present and Future." *Journalism Studies* 6: 543–57.

McNeill, J. R., and Peter Engelke. 2014. *The Great Acceleration: An Environmental History of the Anthropocene since 1945.* Cambridge, Mass.: Harvard University Press.

Merchant, Carolyn. 1989 (1980). *The Death of Nature: Women, Ecology, and the Scientific Revolution.* New York: HarperOne.

Merry, Sally Engle. 2006. *Human Rights and Gender Violence: Translating International Law into Local Justice*. Chicago: University of Chicago Press.

———. 2011. "Measuring the World: Indicators, Human Rights and Global Governance." *Current Anthropology* 52: S83–S95.

———. 2016. *The Seductions of Quantification: Measuring Human Rights, Gender Violence and Sex Trafficking*. Chicago: University of Chicago Press.

Mintz, Sidney. 1985. *Sweetness and Power: The Place of Sugar in Modern History*. New York: Penguin Books.

Mitchell, Timothy. 1998. "Fixing the Economy." *Cultural Studies* 12(1): 82–101.

———. 2002. *Rule of Experts: Egypt, Techno-Politics, Modernity*. Berkeley: University of California Press.

———. 2006. "Rethinking Economy." *Geoforum* 39: 1116–21.

Miyake, M., P. Guillotreau, C.-H. Sun, and G. Ishimura. 2010. "Recent Developments in the Tuna Industry: Stocks, Fisheries, Management, Processing, Trade and Markets." FAO Fisheries and Aquaculture Technical Paper No. 543. Rome: Food and Agricultural Organization of the United Nations.

Montgomery, Sy. 2015. *The Soul of an Octopus: A Surprising Exploration into the Wonder of Consciousness*. New York: Atria Books.

Moore, Jason W. 2015. *Capitalism in the Web of Life: Ecology and the Accumulation of Capital*. New York: Verso.

Moore, Sally Falk. 1978. *Law as Process: An Anthropological Approach*. Boston: Routledge.

Myers, Fred, ed. 2001. *The Empire of Things: Regimes of Value and Material Culture*. Santa Fe, N.Mex.: School of American Research Press.

Myers, Ransom A., and Boris Worm. 2003. "Rapid Worldwide Depletion of Predatory Fish Communities." *Nature* 423: 280–83.

Nadelson, Robert. 1992. "The Exclusive Economic Zone: State Claims and the LOS Convention." *Marine Policy* 16: 463–87.

Nader, Laura. 1996. *Naked Science: Anthropological Inquiry into Boundaries, Power and Knowledge*. New York: Routledge.

Nandan, S. N. 1987. "The Exclusive Economic Zone: A Historical Perspective." *FAO Essays in Memory of Jean Carroz. The Law and the Sea*. Rome: FAO. Available online at http://www.fao.org/docrep/s5280T/s5280t0p.htm.

Nappi, Carmine. 1979. *Commodity Market Controls: A Historical Review*. Lexington, Mass.: Lexington Books.

Nealon, Jeffrey T. 2016. "The Archeology of Biopower: From Plant to Animal Life in *The Order of Things*." In *Biopower: Foucault and Beyond*, ed. Vernon W. Cisney and Nicolae Morar, 138–57. Chicago: University of Chicago Press.

Neumann, Iver B. 2012. *At Home with the Diplomats: Inside a European Foreign Ministry*. Ithaca, N.Y.: Cornell University Press.

Nibert, David. 2002. *Animal Rights and Human Rights: Entanglements of Oppression and Liberation*. New York: Rowman & Littlefield Publishers.

Nixon, Rob. 2011. *Slow Violence and the Environmentalism of the Poor*. Cambridge, Mass.: Harvard University Press.

Olds, Kris, and Nigel Thrift. 2005. "Cultures on the Brink: Reengineering the Soul of Capitalism—On a Global Scale." In *Global Assemblages: Technology, Politics, and Ethics as Anthropological Problems,* ed. Aihwa Ong and Stephan J. Collier, 270–90. Malden, Mass.: Blackwell Publishing.

Ong, Aihwa, and Stephan J. Collier, eds. 2005. *Global Assemblages: Technology, Politics, and Ethics as Anthropological Problems.* Malden, Mass: Blackwell Publishing.

O'Reilly, Jessica. 2017. *The Technocratic Antarctic: An Ethnography of Scientific Expertise and Environmental Governance.* Ithaca, N.Y.: Cornell University Press.

Oreskes, Naomi, and Erik M. Conway. 2010. *Merchants of Doubt: How a Handful of Scientists Obscured the Truth on Issues from Tobacco Smoke to Global Warming.* New York: Bloomsbury.

Ortner, Sherry B. 1996. *Making Gender: The Politics and Erotics of Culture.* Boston: Beacon Press.

———. 2006. *Anthropology and Social Theory: Culture, Power, and the Acting Subject.* Durham, N.C.: Duke University Press.

Ostrom, Elinor, and Edella Schlager. 1996. "The Formation of Property Rights." In *Rights to Nature: Ecological, Economic, Cultural, and Political Principles of Institutions for the Environment,* ed. Susan Hanna, Carl Folke, and Karl-Goren Maler, 127–56. Washington, D.C.: Island Press.

Ostrom, Elinor, Joanna Burger, Christopher B. Field, Richard B. Norgaard, and David Policansky. 1999. "Revisiting the Commons: Local Lessons, Global Challenges." *Science* 284: 278–82.

Pagden, Anthony. 1994. *The Uncertainties of Empire: Essays in Iberian and Ibero-American Intellectual History.* Brookfield, Vt.: Ashgate Publishing.

Parmentier, Remi. 2012. "Role and Impact of International NGOs in Global Ocean Governance." In *Ocean Yearbook 26,* ed. A. Chircop, S. Coffen-Smout, and M. McConnell, 209–30. Boston: Martinus Nijhoff Publishers.

Parreñas, Juno Salazar. 2018. *Decolonizing Extinction: The Work of Care in Orangutan Rehabilitation.* Durham, N.C.: Duke University Press.

Patton, Paul. 2016. "Power and Biopower in Foucault." In *Biopower: Foucault and Beyond,* ed. Vernon W. Cisney and Nicolae Morar, 102–17. Chicago: University of Chicago Press.

Pauly, Daniel. 1995. "Anecdotes and the Shifting Baseline Syndrome of Fisheries." *Trends in Ecology and Evolution* 10: 430.

———. 2009. "Aquacalypse Now. The End of Fish." *New Republic,* 24–27.

———. 2010. *5 Easy Pieces: How Fishing Impacts Marine Ecosystems.* Washington, D.C.: Island Press.

Peterson, M. J. 1993. "International Fisheries Management." In *Institutions for the Earth: Sources of Effective International Environmental Protection,* ed. P. M. Haas, R. O. Keohane, and M. A. Levy, 249–305. Cambridge, Mass.: MIT Press.

Pieribone, Vincent, and David Gruber. 2005. *Aglow in the Dark: The Revolutionary Science of Biofluorescence.* Cambridge, Mass.: Harvard University Press.

Pistor, Katharina. 2019. *The Code of Capital: How the Law Creates Wealth and Inequality.* Princeton, N.J.: Princeton University Press.

Plumwood, Val. 1993. *Feminism and the Mastery of Nature.* New York: Routledge.

Polanyi, Karl. 2001 (1944). *The Great Transformation: The Political and Economic Origins of Our Time.* Boston: Beacon Press.

Poovey, Mary. 1998. *A History of the Modern Fact: Problems of Knowledge in the Sciences of Wealth and Society.* Chicago: University of Chicago Press.

——. 2015. "On 'The Limits to Financialization.'" *Dialogues in Human Geography* 5(2): 220–24.

Porch, Clay. 2005. "The Sustainability of Western Atlantic Bluefin Tuna: A Warm-Blooded Fish in a Hot-Blooded Fishery." *Bulletin of Marine Science* 76: 363–84.

Porter, Theodore M. 1995. *Trust in Numbers: The Pursuit of Objectivity in Science and Public Life.* Princeton, N.J.: Princeton University Press.

Power, Michael. 1999. *The Audit Society: Rituals of Verification.* New York: Oxford University Press.

Powers, Matthew. 2018. *NGOs as Newsmakers: The Changing Landscape of International News.* New York: Columbia University Press.

Probyn, Elspeth. 2016. *Eating the Ocean.* Durham, N.C.: Duke University Press.

Purcell, Nicholas. 1995. "Eating Fish: The Paradoxes of Seafood." In *Food in Antiquity,* ed. John Wilkins, David Harvey, and Michael Dobson, 132–49. Exeter: University of Exeter Press.

——. 2003. "The Way We Used to Eat: Diet, Community, and History at Rome." *American Journal of Philology* 124: 329–58.

Purdy, Jedediah. 2015. *After Nature: A Politics for the Anthropocene.* Cambridge, Mass.: Harvard University Press.

Raffles, Hugh. 2011. *Insectopedia.* New York: Vintage Books.

Rajagopol, Arvind. 2013. "Putting America in Its Place." *Public Culture* 25: 387–99.

Rancière, Jacques. 2004. *The Politics of Aesthetics: The Distribution of the Sensible.* Trans. Gabriel Rockhill. New York: Continuum.

Reinharz, Shulamit. 1992. *Feminist Methods in Social Research.* New York: Oxford University Press.

Rigney, Matt. 2012. *In Pursuit of Giants: One Man's Global Search for the Last of the Great Fish.* Lebanon, N.H.: University Press of New England.

Riles, Annelise. 2011. *Collateral Knowledge: Legal Reasoning in the Global Financial Markets.* Chicago: University of Chicago Press.

Ritvo, Harriet. 1997. *The Platypus and the Mermaid and Other Figments of the Classifying Imagination.* Cambridge, Mass.: Harvard University Press.

Roberts, Callum. 2007. *The Unnatural History of the Sea.* Washington, D.C.: Island Press.

Römer, Johanna. 2019. "Ambiguous Publicities: Cultivating Doubt at the Intersection of Competing Genres of Risk Evaluation in Catalan Prisons." *Punishment & Society* 21(3): 342–63.

Rosaldo, Renato. 1989. "Imperialist Nostalgia." *Representations* 26: 107–22.

Rosenau, James N. 1992. "Governance, Order, and Change in World Politics." In *Governance without Government: Order and Change in World Politics*, ed. James N. Rosenau and Ernst-Otto Czempiel, 1–29. New York: Cambridge University Press.

Ross, Carne. 2007. *Independent Diplomat: Dispatches from an Unaccountable Elite.* Ithaca, N.Y.: Cornell University Press.

Ruais, Rich. 2011/2012. "Requiem for 'Iconic Bluefin' Extinction Myth: A Seven-Part Series by ABTA [Atlantic Bluefin Tuna Association] Executive Director Rich Ruais." *Commercial Fisheries News* 1–22. https://keper-connell .squarespace.com/s/CFN_8_11-2_12.pdf.

Safina, Carl. 1993. "Bluefin Tuna in the West Atlantic: Negligent Management and the Making of an Endangered Species." *Conservation Biology* 7: 229–34.

———. 1997. *Song for the Blue Ocean: Encounters along the World's Coasts and Beneath the Seas.* New York: Henry Holt.

———. 2001. "Tuna Conservation." In *Tuna: Physiology, Ecology, and Evolution*, ed. Barbara A. Block and E. Donald Stevens, 413–59. New York: Academic Press.

Sara, R. 1980. "Bluefin Tuna Trap Fishing in the Mediterranean." Collective Volume of Scientific Papers ICCAT 11: 129–44.

Scheper-Hughes, Nancy. 1995. "The Primacy of the Ethical: Propositions for a Militant Anthropology." *Current Anthropology* 36(3): 409–40.

Schmitt, Carl. 1997 (1942). *Land and Sea.* Trans. Simona Draghici. Washington, D.C.: Plutarch Press.

Scott, James C. 1998. *Seeing like a State: How Certain Schemes to Improve the Human Condition Have Failed.* New Haven, Conn.: Yale University Press.

Sharp, Gary D. 2001. "Tuna Oceanography—An Applied Science." In *Tuna: Physiology, Ecology, and Evolution*, ed. Barbara A. Block and E. Donald Stevens, 345–89. New York: Academic Press.

Shore, Cris, and Susan Wright. 1997. *Anthropology of Policy: Critical Perspectives on Governance and Power.* New York: Routledge.

Shore, Cris, Susan Wright, and Davide Però, eds. 2011. *Policy Worlds: Anthropology and the Analysis of Contemporary Power.* New York: Berghahn Books.

Shukin, Nicole. 2009. *Animal Capital: Rendering Life in Biopolitical Times.* Minneapolis: University of Minnesota Press.

Silbey, Susan S., and Patricia Ewick. 2003. "The Architecture of Authority: The Place of Law in the Space of Science." In *The Place of Law*, ed. Austin Sarat, Lawrence Douglas, and Martha Merrill Umphrey, 75–108. Ann Arbor: University of Michigan Press.

Sinclair, Michael, and Per Solemdal. 1988. "The Development of 'Population Thinking' in Fisheries Biology between 1878 and 1930." *Aquatic Living Resources* 1(3): 189–213.

Singh-Renton, S. 2010. "Sustainable Development and Conservation of Tuna and Tuna-Like Species in the Caribbean: The Role of ICCAT." CARICOM. CRFM Technical & Advisory Document Series. Number 2010/2.

Slaughter, Anne-Marie. 2004. *A New World Order*. Princeton, N.J.: Princeton University Press.

Sloan, Stephan. 2003. *Ocean Bankruptcy: World Fisheries on the Brink of Disaster*. Guilford, Conn.: Lyons Press.

Smith, Tim D. 1994. *Scaling Fisheries: The Science of Measuring the Effects of Fishing, 1955–1995*. New York: Cambridge University Press.

Smuts, Barbara. 2001. "Encounters with Animal Minds." *Journal of Consciousness Studies* 8: 293–309.

Spates, James L. 1983. "The Sociology of Values." *Annual Review of Sociology* 9: 27–49.

Starosielski, Nicole. 2015. *The Undersea Network*. Durham, N.C.: Duke University Press.

Steffen, Will, Jacques Grinevald, Paul Crutzen, and John McNeill. 2011. "The Anthropocene: Conceptual and Historical Perspectives." *Philosophical Transactions of the Royal Society A* 369: 842–67.

Stengers, Isabelle. 2018. *Another Science Is Possible: A Manifesto for Slow Science*. Trans. Stephen Muecke. Medford, Mass.: Polity.

Stolba, Vladimir F. 2005. "Fish and Money: Numismatic Evidence for Black Sea Fishing." In *Ancient Fishing and Fish Processing in the Black Sea Region*, ed. T. Bekker-Nielsen, 115–32. Oakville, Conn.: Aarhus University Press.

Strathern, Marilyn. 2000. *Audit Cultures: Anthropological Studies in Accountability, Ethics and the Academy*. New York: Routledge.

Suárez de Vivero, Juan Luís, and Juan Carlos Rodríguez Mateos. 2002. "Spain and the Sea: The Decline of an Ideology, Crisis in the Maritime Sector and the Challenges of Globalization." *Marine Policy* 26: 143–53.

Sullivan, Sian. 2017. "On 'Natural Capital,' 'Fairy Tales' and Ideology." *Development and Change* 48(2): 397–423.

Tamanaha, Brian Z. 2006. *Law as a Means to an End: Threat to the Rule of Law*. New York: Cambridge University Press.

Telesca, Jennifer E. 2015. "Consensus for Whom?: Gaming the Market for Atlantic Bluefin Tuna through the Empire of Bureaucracy." *Cambridge Journal of Anthropology* 33(1): 49–64.

——. 2017. "Accounting for Loss in Fish *Stocks*: A Word on Life as Biological Asset." *Environment and Society: Advances in Research* 8: 144–60.

——. 2018. "Volatility." *Cultural Anthropology* from the series "Speaking Volumes," ed. Franck Billé. https://culanth.org/fieldsights/1464-volatility.

——. Forthcoming. "Fishing for the Anthropocene: Time in Ocean Governance." In *Timescales: Ecological Temporalities Across Disciplines*, ed. Carolyn Fornoff, Patricia Eunji Kim, and Bethany Wiggin. Minneapolis: University of Minnesota Press.

Ticktin, Miriam. 2017. "A World without Innocence." *American Ethnologist* 44(4): 577–90.

Tsing, Anna Lowenhaupt. 2005. *Friction: An Ethnography of Global Connection*. Princeton, N.J.: Princeton University Press.

———. 2012. "Unruly Edges: Mushrooms as Companion Species." *Environmental Humanities* 1: 141–54.

———. 2015. *The Mushroom at the End of the World: On the Possibility of Life in Capitalist Ruins*. Princeton, N.J.: Princeton University Press.

———. 2016. "Earth Stalked by Man." *Cambridge Journal of Anthropology* 34: 2–16.

Tsing, Anna, Heather Swanson, Elaine Gan, and Nils Bubandt. 2017. *Arts of Living on a Damaged Planet*. Minneapolis: University of Minnesota Press.

Turem, Ziya Umut, and Andrea Ballestero. 2014. "Regulatory Translations: Expertise and Affect in Global Legal Fields." *Indiana Journal of Global Legal Studies* 21(1): 1–25.

Turk, Austin T. 1976. "Law as a Weapon in Social Conflict." *Social Problems* 23(3): 276–91.

Turner, Victor. 1967.*The Forest of Symbols: Aspects of Ndembu Ritual*. Ithaca, N.Y.: Cornell University Press.

Urbina, Ian. 2019. *The Outlaw Ocean: Journeys across the Last Untamed Frontier*. New York: Alfred A. Knopf.

Van Dooren, Thom. 2014. *Flight Ways: Life and Loss at the Edge of Extinction*. New York: Columbia University Press.

Varoufakis, Yanis. 2017. *Adults in the Room: My Battle with the European and American Deep Establishment*. New York: Farrar, Straus and Giroux.

Venzke, Ingo. 2017. "Possibilities of the Past: Histories of the NIEO and the Travails of Critique." ACIL Research Paper 2017–26.

Verdery, Katherine. 2018. *My Life as a Spy: Investigations in a Secret Police File*. Durham, N.C.: Duke University Press.

Vine, David. 2009. *Island of Shame: The Secret History of the U.S. Military Base on Diego Garcia*. Princeton, N.J.: Princeton University Press.

Wang, Chen. 2010. "Issues on Consensus and Quorum at International Conferences." *Chinese Journal of International Law* 9: 717–39.

Warner, Michael. 2002. *Publics and Counterpublics*. New York: Zone Books.

Weber, Max. 1964 (1947). *The Theory of Social and Economic Organization*. Ed. Talcott Parsons. New York: Free Press.

———. 1978 (1922). *Economy and Society: An Outline of Interpretive Sociology*. Berkeley: University of California Press.

Webster, D. G. 2008. *Adaptive Governance: The Dynamics of Atlantic Fisheries Management*. Cambridge, Mass.: MIT Press.

Wedel, Janine R. 2009. *Shadow Elite: How the World's New Power Brokers Undermine Democracy, Government and the Free Market*. New York: Basic Books.

Whynott, Douglas. 1995. *Giant Bluefin*. New York: Farrar, Straus and Giroux.

Wiber, Melanie G. 2005. "Mobile Law and Globalism: Epistemic Communities versus Community-Based Innovation in the Fisheries Sector." In *Mobile People, Mobile Law: Expanding Legal Relations in a Contracting World*, ed. F. von Benda-Beckmann, K. von Benda-Beckmann, and A. Griffiths, 131–52. Burlington, Vt.: Ashgate Publishing.

Williams, Raymond. 1977. *Problems in Materialism and Culture: Selected Essays*. New York: Verso.

Wilson, Richard Ashby. 2001. *The Politics of Truth and Reconciliation in South Africa: Legitimizing the Post-Apartheid State*. New York: Cambridge University Press.

———. 2011. *Writing History in International Criminal Trials*. New York: Cambridge University Press.

Woolf, Stuart. 1989. "Statistics and the Modern State." *Comparative Studies in Society and History* 31(3): 588–604.

Wosnitzer, Robert. 2016. "Mapping the Trading Desk: Derivative Value through Market Making." In *Derivatives and the Wealth of Societies*, ed. Benjamin Lee and Randy Martin, 252–74. Chicago: University of Chicago Press.

Wynter, Sylvia. 1994. "'No Humans Involved': An Open Letter to My Colleagues." In *Forum N.H.I.: Knowledge for the 21st Century*, 42–71. Stanford, Calif.: Institute N.H.1.

Yusoff, Kathryn. 2018. *A Billion Black Anthropocenes or None*. Minneapolis: University of Minnesota Press.

Index

~~~

Page references in italics refer to figures.

Anthropocene and, xxiii, 3; conditions for, 56; MSY and, 53
Greece, tuna farms (ranches) and, 94
Greenberg, Paul, 122, 134
Greenpeace, 106, 115, 128, 136, 195, 264n48; noncompliance and, 265n61; protest by, 191
Greenpeace France, 123, 123–24, 191; protest by, 192
Greenpeace Spain, 121
Grotius, Hugo, 86
Guggenheim Museum (Bilbao), 198
Gulf of Mexico, 36, 69, 174, 188, 213; MPAs in, 196; spawning in, 89

haddock, 115, 145
Haraway, Donna, 6, 29, 223
Hardin, Garret, 100, 261n90. See also tragedy of the commons
Hardt, Michael, 38, 40–41, 64, 102, 136, 142
Harrods Department Store, 84
harvesting, 34, 51, 170, 221
Hemingway, Ernest, xx
history, 33, 98, 147, 158; imperial, 72; ocean and, 34–36
Hoefnagel, Georg, 254n61
Hogarth, William, 265n60
Hogenberg, Franz, 254n61
Holling, C. S., 52
Holocene, 3, 246n10
Horden, Peregrine, 43
"How the Illegal Bluefin Tuna Market Made over EUR 12 Million a Year Selling Fish in Spain" (EUROPOL), 133
Huizinga, John, 197
human rights, 105, 112, 267n39
Human Rights Watch, 112
Huxley, Thomas Henry, 49
Hviding, Edvard, 109

IATTC. See Inter-American Tropical Tuna Commission
ICCAT. See International Commission

for the Conservation of Atlantic Tunas
"ICCAT Criteria for the Allocation of Fishing Possibilities" (document 01-25), 98
Iceland, 36, 236; Libyan proposal and, 187, 190; quotas and, 238
illegal, unreported, and unregulated (IUU) fishing, xvii, 83, 96, 118, 126–27, 132, 133, 164, 172, 188, 191, 266n65
imperialism, 37, 64, 70, 73, 224, 252n16
Inconvenient Truth, An (film), 120–21
Indian Ocean Tuna Commission (IOTC), 249n66, 256n124
industrial fishers, 18, 77, 107, 121
Industrial Revolution, 246n10
Institute for Public Knowledge (IPK), 24, 263n24
Inter-American Tropical Tuna Commission (IATTC), 248n66, 258n7, 267n38
intergovernmental organizations, 6, 16, 19–20, 32, 105, 176, 179, 184, 206
Intergovernmental Panel on Climate Change (IPCC), 154
Intergovernmental Science-Policy Platform on Biodiversity and Ecosystem Services (IPBES), 245n9
International Chamber of Shipping, 258n161
International Commission for the Conservation of Atlantic Tunas (ICCAT), 33, 37, 57, 68, 72, 91, 154, 177, 197, 218; agenda of, 110–11; audit culture of, 141; bureaucratic survival of, 163; challenging, xxiii, 115, 116; commodity empires and, 7, 117, 207–8, 217; criticism of, 108, 113, 124, 139, 204–5, 217–18; economic growth and, 4; export markets and,

International Union for Conservation of Nature (IUCN), xvii, 244n16
International Whaling Commission (IWC), 113, 179, 181, 248n63, 263n27
IOTC. *See* Indian Ocean Tuna Commission
IPK. *See* Institute for Public Knowledge
Issenberg, Sasha, 32, 59
Istanbul, meeting in, 127, 163, 198
Italy, xviii, 55, 96, 114, 121, 236, 256n125, 257n143; quotas for, 98, 238, 239
IUU fishing. *See* illegal, unreported, and unregulated fishing
IWC. *See* International Whaling Commission

JAL (Japan Airlines), 59, 60–61, 62
Japan, xvi, 16, 19, 20, 24, 37, 39, 78, 116, 195, 197, 209, 210, 235, 249n84, 255n87, 270n31; certificate of origin and, 117; CITES and, 117, 122, 186, 187, 204; constraints on, 93; control by, 184; EEZs and, 91, 259n38; fisheries science and, 52, 93, 134, 152, 153, 155, 169, 171; foreign aid from, 11; ICCAT's founding and, 54–57, 256n125; industrial fishing fleet of, xviii, 47, 77, 86, 91, 114; Libyan proposal and, 187–92; management measures and, 89; *New York Times* and, 128–29; NGOs, 136; proposals by, 172, 202–4; quotas and, 84, 89–91, 97, 98, 99, 115, 122, 172, 202, 237, 238; sushi economy and, 57–63, 257n143, 257n151; U.S. occupation of, 47–50, 59; voluntary measures and, 90–91; World War II and, 46–47, 53, 63, 257n142
Japanese Ministry of Finance, 61, 255n80, 257n155
jellyfish, 29

Jolly, David, 120, 121–22, 124, 126, 127, 128

knowledge: hierarchies of, 144–54; indigenous/traditional, 273n37; production of, 155, 156, 173, 267n44; representation of, 194; scientific, 144, 147, 149, 154
Kobe Matrix, 169, 170, 171, 216
Kopytoff, Igor, xvi
Korea, 47, 235, 256n125; Libyan proposal and, 187, 190; quotas and, 97, 201, 239
Koskenniemi, Martti, 74, 76

labor: attack on, 58; costs, 58; developments in, 63, 70; standards, 13
Larkin, P. A., 50, 52
law, 30, 176, 217, 220; administrative, 68, 100, 219, 248n58; commodity empires and, 48–53; fisheries science and, 173; formations of, 73; power of, 84, 177, 207; as process, 70, 71, 72–79, 179–80; regularization and, 74–75, 84, 98; regulatory, 89; situational adjustment and, 74–75, 179–80, 184, 203; statutory, 100; uncertainty and, 74–75; vector of, 37, 72, 99, 118. *See also* international law
Law of the Sea Conventions. *See* United Nations Convention on the Law of the Sea
Lee, Benjamin, 61
LePuma, Edward, 61
Levinson, Marc, 63, 258n161
Libya, 140, 155, 186, 206, 235, 269n21; moratorium proposed, 90; play on rules by, 177, 183–92; proposals by, 185, 187–88, 189, 191; quotas and, 96, 97, 200, 203, 239; strategy by, 184, 203, 204
life-forms, 2, 5, 15, 79; monetized, 37, 38

MSY. *See* maximum sustainable yield

multilateralism, 219, 221

Murphy, Dennis, 262n1

Nadelson, Robert, 79

narratives, 125–28; dominant, 108–9; framing, 105–10

National Audubon Society, 106, 115, 116, 264n39

National Coalition for Marine Conservation, 264n39

National Geographic, 157

nationalism, 207, 252n20; "resource," 37, 38, 86, 194–95

National Oceanic and Atmospheric Administration (NOAA), 67, 184

National Resources Defense Council (NRDC), 262n2

National Swedish Board of Fisheries, 116

nation-states, 2, 9, 11, 16, 34, 45, 49, 51, 71, 81–82, 200; compliance by, 78; economic losses for, 93; extermination and, 39; ICCAT and, 24, 40, 77–78; MSY and, 52; natural "resources" and, 99

NATO, 185

Nature, xii, 6, 11, 27, 28, 103, 105, 133, 216; abstract, 135; civilized and, 214–15; compartmentalization of, 135; humanizing, 154–60

nature, 15, 21, 34, 47, 102, 145, 150, 159, 215; alienation of, 223; capitalism and, 8; culture and, 137; human activity and, 223

Negri, Antonio, 38, 40–41, 64, 102, 136, 142

neoliberalism, 64, 92, 119, 253n25. *See also* geoeconomics

Neumann, Iver, 74, 218, 251n119

New International Economic Order (NIEO), 260n39

New York Times, 28, 110, 114, 116, 129, 137, 236n21, 236n23; circulation of, 266n67; eastern Atlantic bluefin

campaign coverage by, 121–28, 132; editorializing the news and, 119–21; ICCAT and, 111–13, 119, 121–22, 125–26, 127–28, 132–33, 134, 189, 263n21; as newspaper of record, 107, 111–13; NGOs and, 112; patterns in analyzing, 263n21; political economy and, 108, 113, 263n13; United Nations and, 112; western Atlantic bluefin campaign coverage by, 113–18, 121, 132. *See also* savior plot

New York Times Sunday Magazine, 133, 134

NGOs. *See* nongovernmental organizations

Nixon, Richard M., 61, 86, 95

"Nixon shock," 61, 62

NOAA. *See* National Oceanic and Atmospheric Administration

Nobu restaurants, 94

nongovernmental organizations (NGOs), 104, 105, 108, 121, 127, 185; appearance of, 128; biopolitics and, 136; communications policy of, 131; diversification of, 119; environmental, 76, 106, 112–13, 135, 152, 175, 183, 195, 263n24; environmentalists and, 136, 186; growth of, 112, 119; professionalization of, 119, 128; strategizing by, 111–12

Norway, 36, 77, 144, 145, 155, 236, 239, 261n79, 271n47; decision-making process and, 204–5; fisheries management and, 144–45; Libyan proposal and, 187, 190

Norwegian Ministry of Foreign Affairs, 74

Nyman, Lennert, 116

Obama, Barack, 132, 184

Oceana, 68, 106, 136, 195

ocean acidification, 3, 149, 162, 164, 213

Stengers, Isabelle, 219
"stock" assessments, 29, 96, 125, 152, 153, 154, 171, 192; uncertainty in, 160–68, 169
"stocks," 18, 27, 48, 70, 71, 77, 78, 90, 101, 107, 110–11, 133, 143, 144, 160, 165, 169, 173, 183, 202, 210; depletion of, 56, 89, 108, 162, 188; future of, 31, 64; growth of, 160–61, 162, 221; ICCAT and, 108, 194, 195; mixing of, 89; size of, 162; splitting, 83–94; truths about, 156
Strabo, 43
Strait of Gibraltar, xv
Sunoco, 136
supranationality, 17, 20, 23, 71, 182
sushi economy, xvi, xix, 1, 4, 82, 95, 178, 209; birth of, 57–63; sashimi market and, 257n143; tuna farms (ranches) and, 94
Sushi Economy, The (Issenberg), 59
sustainability, 4, 157, 184
Suzuki, Ziro, 134
Sweden, 116, 117
swordfish, xvii, xx, 1, 18, 82, 97, 103, 212, 262n2. See also billfish

TAC. See total allowable catch
taxonomies, xv, 32, 248n58
technocrats, 29, 50, 164, 216, 218; mass extinction and, 222–23
techno-fixes, 127, 188, 220
technology, 20, 33, 46, 62–63, 173, 215, 218, 219, 220, 223; development and, 50; digital, 119; fisheries science and, 159; underwater video, 157
techno-speak, 153, 154, 159
Tertiary Period, 35
Thatcher, Margaret, 92
Théâtre du Merveilleux, dinner at, 198, 199
Thrift, Nigel, 12
Tokyo Sea Life Park, 210, 211; photo of, 212. See also aquarium

total allowable catch (TAC), 83, 96, 144, 149, 158, 160, 171, 200; proposed, 202–3; sharp increase in, 132–33. See also two-stock theory
trade, 16, 18; global commodities, 33, 72, 180; global rules of, 13, 79; imbalance, 61; joust over, 62; rules, 78, 97; terms of, 73
tragedy of the commons, xvii, 28, 72, 99–101, 122, 262n96
trap fishery, xviii–xix, 44–46, 53, 58, 98, 158, 254n61
Truman, Harry, 80, 95
Truman Proclamation, 80
Tsing, Anna, 42, 61, 70, 215, 252n8
Tsukiji marketplace, xvi, 59, 61, 95, 104; bluefin tuna trade and, 62, 94; photo of, 210; tuna auction at, xvi, 94, 115, 209, 210, 257n143; visiting, 209
Tudela, Sergi, 121
tuna auction. See Tsukiji marketplace
tuna farms (ranches), xix, 64, 65, 121, 157, 195, 211, 215, 216; development of, 94; rules for, 96
"Tuna's End" (New York Time Sunday Magazine), 133, 134
Tunisia, 130, 195, 235; decision-making process and, 204; Libyan proposal and, 187, 190; quotas and, 97, 237
tunny, 43, 250n91
Turk, Austin, 73
Turkey, 21, 42, 94, 130, 155, 236, 269n17, 271n48; decision-making process and, 200, 204; Libyan proposal and, 187, 190; quotas and, 99, 172, 203, 204, 239; tuna farms (ranches) and, 94
two-stock theory, 83, 84, 87–88, 88, 88–89, 90

uncertainty, 37, 172, 195; in economy, 13; in fisheries management, 169, 171, 172; in fisheries science, 141,

World Trade Organization (WTO), 10, 118, 205

World War II, 9, 34, 48, 53, 59, 63, 64, 71, 79, 80, 85, 181; fish decline since, 5; fishing effort after, xxii, 244n37; food shortages during, 47; impact of, 46, 47; industrialization since, 3; overexploitation since, 211. *See also* Great Acceleration

World Wide Fund for Nature (WWF), 106, 116, 121, 136, 195, 264n39, 264n48; lobbying by, 264n36; noncompliance and, 128–31, 265n61

WTO. *See* World Trade Organization

WWF. *See* World Wide Fund for Nature

Wynter, Sylvia, 243n1

yellowfin tuna, xv, xviii, 18, 47, 58, 103, 162, 211, 258n7

Jennifer E. Telesca is assistant professor of environmental justice in the Department of Social Science and Cultural Studies at the Pratt Institute.